AF598042

METHODS IN MOLECULAR BIOLOGY™

Series Editor
John M. Walker
School of Life Sciences
University of Hertfordshire
Hatfield, Hertfordshire, AL10 9AB, UK

For other titles published in this series, go to
www.springer.com/series/7651

RNAi and Plant Gene Function Analysis

Methods and Protocols

Edited by

Hiroaki Kodama

Chiba University, Chiba, Japan

Atsushi Komamine

The Research Institute of Evolutionary Biology,
Tokyo, Japan

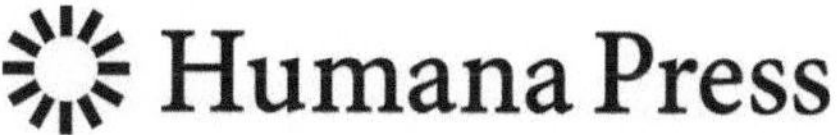

Editors
Hiroaki Kodama
Graduate School of Horticulture
Chiba University
Yayoi-cho 1-33
263-8522 Chiba, Inage-ku, Japan
kodama@faculty.chiba-u.jp

Atsushi Komamine
The Research Institute of Evolutionary Biology
Kamiyoga 2-4-28
158-0098 Tokyo, Setagaya, Japan
khf10654@nifty.com

ISSN 1064-3745
e-ISSN 1940-6029
ISBN 978-1-61779-122-2
e-ISBN 978-1-61779-123-9
DOI 10.1007/978-1-61779-123-9
Springer New York Dordrecht Heidelberg London

Library of Congress Control Number: 2011926633

Printed on acid-free paper

Humana Press is part of Springer Science+Business Media (www.springer.com)

Preface

Recent rapid and unprecedented development of DNA sequencing technology has enabled us to obtain the entire genomic information with extremely low costs within a very short period. As sequence information increases, the need for functional annotation of target genes also increases. Due to the low efficiency of homologous recombination in plants, targeted gene destruction has not been used as a way to knock down gene function.

RNA interference (RNAi) was discovered by Fire, Mello and their colleagues in 1998 and its use has rapidly expanded into the broad area of life science, especially animal and plant cell studies. RNAi, which can ectopically reduce the expression of target genes, is guided by 21–25-nucleotide-long small RNAs (sRNAs). This key molecule was discovered in plants by Hamilton and Baulcombe in 1998. The double-stranded RNA-specific endonuclease, Dicer, processes double-stranded RNA molecules and sRNAs are produced. A single-stranded sRNA is incorporated into Argonaute (AGO) proteins. mRNAs are cleaved at the middle of the complementary region which has been annealed with an sRNA molecule.

Plant cells are resistant to the direct transfer of sRNAs because of their rigid cell walls. Thus in most cases, RNAi is established after stable transformation of plant cells, and an expression cassette which transcribes an inverted repeat sequence is introduced. Its resulting transcripts form a stem–loop structure (the so-called hairpin RNAs). The stem is processed by Dicer, and then sRNAs are produced. Several bypasses without the production of stable transformants are developed in order to assess the effects of RNAi, and some of them are also described in this book. This volume is intended to guide basic RNAi technologies which have been developed in and for plant biology.

A historical overview of RNAi is provided in **Chapter 1**, in which an attractive story about the discovery and early establishment of the RNAi pathway is described and would be useful for increasing readers' interest. Plant science played an important role at the dawn of RNAi research. In **Chapter 2**, the side effect of RNAi technology, an off-target effect, is reviewed. Although RNAi is certainly a powerful tool for molecular biology, understanding regarding the sequence specificity of RNAi is a pre-requisite for the elucidation of RNAi's effects. Construction of an RNAi vector, especially of an inverted repeat structure, is a difficult and time-consuming process. The use of Gateway technology can markedly reduce this laborious work as introduced in **Chapter 3**. Application of RNAi technology to the analysis of essential genes would be difficult because knockdown of essential genes often causes lethality. In **Chapter 4**, an inducible RNAi vector is introduced, which can control gene expression at the spatial and/or temporal level and circumvent viability problems. An example for artificial microRNA (amiRNA) technology in the moss is introduced in **Chapter 5**. AmiRNA technology was developed recently, in which only one effective sRNA molecule is produced. This method has several advantages in comparison with traditional RNAi vectors. In **Chapter 6**, virus-induced gene silencing (VIGS) is introduced. Infection of plant viruses is often associated with the generation of sRNAs targeting viral sequences. When the plant gene sequences are incorporated into the viral genome, sRNAs harbouring these plant gene sequences are generated after infection of recombinant viruses. In **Chapter 7**, local induction of RNAi by agroinfiltration is introduced. The RNAi vector can be transiently delivered into leaf cells by *Agrobacterium*. A wide-range host of *Agrobacterium* allows us to induce local silencing of target genes in a variety of plants. Direct transfer of sRNAs into protoplasts is also effective for the knockdown of target gene function as is explained in **Chapter 8**. To evaluate the effects of RNAi, researchers often need to detect sRNAs. Detection of sRNAs by northern blot analysis, quantification of sRNAs by qPCR and sequencing of sRNAs at a large scale are introduced in **Chapters 9**, **10** and **11**, respectively. The endogenous sRNAs, especially the so-called microRNAs (miRNAs), are involved in the many aspects of plant development. The prediction of target mRNA is important in functional miRNA analysis. **Chapter 12** provides us with detailed instructions on the computational prediction of miRNA targets. The ensuing four chapters provide instructions on how to evaluate the effects of RNAi. Introduction of RNAi vectors often causes the de novo DNA methylation on sites corresponding to the sRNA sequences (RNA-directed DNA methylation, RdDM). The precise map of methylated cytosine residues can be prepared by bisulphite sequencing of the target genomic regions as described in **Chapter 13**. Induction of DNA methylation at the promoter sequences can reduce transcriptional activity. We can assess the effect of RdDM on transcriptional activity by a nuclear run-on assay as described in **Chapter 14**. Finally, the effects of RNAi can be seen through changes in proteomic profiles. The last two chapters, **Chapters 15** and **16**, provide the instructions for comparative proteomic analysis.

The use of RNAi technology is essential for most plant science researchers. In addition to the functional annotation of unknown genes, RNAi technology has been applied to the genetic engineering of important plant metabolites including starches, oils and storage proteins. RNAi has been used to engineer plants resistant to plant viruses and also to nematodes and insects. This volume will provide many tips on the design of experiments to explore plant gene function to post graduate students and their tutors involved in plant biotechnology and breeding research.

We thank all contributors for their thorough cooperation in the preparation of manuscripts. We also express our gratitude to Springer Science + Business for giving us this opportunity to bring out this book.

Hiroaki Kodama and Atsushi Komamine

Contents

Contributors

TOMOYA ASANO • *Division of Functional Genomics, Advanced Science Research Center, Kanazawa University, Kanazawa, Japan*

M. ASIF ARIF • *Plant Biotechnology, Faculty of Biology, University of Freiburg, Freiburg, Germany*

JEN-CHIH CHEN • *Institute of Biotechnology, National Taiwan University, Taipei, Taiwan*

VINCENT L. CHIANG • *Forest Biotechnology Group, Department of Forestry and Environmental Resources, North Carolina State University, Raleigh, NC, USA*

ELENA DADAMI • *Institute of Molecular Biology and Biotechnology, Heraklion, Crete, Greece; Department of Biology, University of Crete, Heraklion, Crete, Greece*

WILLIAM P. DONOVAN • *Monsanto Company, Chesterfield, MO, USA*

ISAM FATTASH • *Plant Biotechnology, Faculty of Biology, University of Freiburg, Freiburg, Germany*

WOLFGANG FRANK • *Plant Biotechnology, Faculty of Biology, University of Freiburg, Freiburg, Germany*

TOSHIYUKI FUKUHARA • *Department of Applied Biological Sciences, Tokyo University of Agriculture and Technology, Tokyo, Japan*

JEAN-PHILIPPE GALAUD • *UMR 5546 CNRS-Université Toulouse III, Pôle de Biotechnologie végétale, Castanet-Tolosan, France*

NAZMUL HAQUE • *Faculty of Agriculture, Ehime University, Matsuyama, Japan; Stowers Institute for Medical Research, Kansas City, MO, USA*

ROGER P. HELLENS • *The New Zealand Institute for Plant and Food Research, Mt. Albert Research Centre, Auckland, New Zealand*

JUTTA MARIA HELM • *Institute of Molecular Biology and Biotechnology, Heraklion, Crete, Greece; Institute of Applied Genetics and Cell Biology, University of Natural Resources and Applied Life Sciences, Vienna, Austria*

MIYA D. HOWELL • *Monsanto Company, Chesterfield, MO, USA*

CAI-ZHONG JIANG • *Crops Pathology and Genetics Research Unit, USDA-ARS, Davis, CA, USA*

HA-IL JUNG • *Department of Crop and Soil Sciences, Cornell University, Ithaca, NY, USA*

KRITON KALANTIDIS • *Institute of Molecular Biology and Biotechnology, Heraklion, Crete, Greece; Department of Biology, University of Crete, Heraklion Crete, Greece*

BASEL KHRAIWESH • *Department of Plant Systems Biology, Flanders Institute for Biotechnology (VIB), Ghent University, Ghent, Belgium; Department of Plant Biotechnology and Genetics, Ghent University, Ghent, Belgium; Center for Plant Stress Genomics and Technology, 4700 King Abdullah University of Science and Technology, Thuwal, Kingdom of Saudi Arabia*

ERI KIYOTA • *Department of Applied Biological Sciences, Tokyo University of Agriculture and Technology, Tokyo, Japan*

KYUN OH LEE • *Division of Applied Life Science (BK21 Program) and PMBBRC, Gyeongsang National University, Jinju, Korea*
SANG YEOL LEE • *Division of Applied Life Science (BK21 Program) and PMBBRC, Gyeongsang National University, Jinju, Korea*
SHANFA LU • *Chinese Academy of Medical Sciences and Peking Union Medical College, Institute of Medicinal Plant Development, Medicinal Plant Cultivation Research Center, Beijing, China*
FRÉDÉRIC MASCLAUX • *Department of Plant Molecular Biology, University of Lausanne, Lausanne, Switzerland*
HIROMITSU MORIYAMA • *Department of Applied Biological Sciences, Tokyo University of Agriculture and Technology, Tokyo, Japan*
TOSHIYA MURANAKA • *Department of Biotechnology, Graduate School of Engineering, Osaka University, Suita, Osaka, Japan*
KIRANKUMAR S. MYSORE • *Plant Biology Division, The Samuel Roberts Noble Foundation, Ardmore, OK, USA*
MASAMICHI NISHIGUCHI • *Faculty of Agriculture, Ehime University, Matsuyama, Japan*
TAKUMI NISHIUCHI • *Division of Functional Genomics, Advanced Science Research Center, Kanazawa University, Kanazawa, Japan*
RYO OKADA • *Department of Applied Biological Sciences, Tokyo University of Agriculture and Technology, Tokyo, Japan*
MICHAEL REID • *Department of Plant Sciences, University of California, Davis, CA, USA*
MUTHAPPA SENTHIL-KUMAR • *Plant Biology Division, The Samuel Roberts Noble Foundation, Ardmore, OK, USA*
RUI SHI • *Forest Biotechnology Group, Department of Forestry and Environmental Resources, North Carolina State University, Raleigh, NC, USA*
YING-HSUAN SUN • *Forest Biotechnology Group, Department of Forestry and Environmental Resources, North Carolina State University, Raleigh, NC, USA*
SYUNICHI URAYAMA • *Department of Applied Biological Sciences, Tokyo University of Agriculture and Technology, Tokyo, Japan*
ERIKA VARKONYI-GASIC • *The New Zealand Institute for Plant and Food Research, Mt. Albert Research Centre, Auckland, New Zealand*
OLENA K. VATAMANIUK • *Department of Crop and Soil Sciences, Cornell University, Ithaca, NY, USA*
YUICHIRO WATANABE • *Department of Life Sciences, University of Tokyo, Tokyo, Japan*
ZHIYANG ZHAI • *Department of Crop and Soil Sciences, Cornell University, Ithaca, NY, USA*
YUANJI ZHANG • *Monsanto Company, Chesterfield, MO, USA*

Chapter 1

Overview of Plant RNAi

Yuichiro Watanabe

Abstract

In the last decade, much progress has been made towards a basic understanding of RNA silencing mechanisms in plants, like in animals and other eukaryotes. Many events that were already known, such as pathogen-derived resistance, posttranscriptional gene silencing, and microRNA (miRNA)-mediated regulation, were found to share a fundamentally similar mechanism. By taking advantage of such mechanisms, whether deliberately or not, we can suppress some biological activities by targeting a specific gene function. This type of applied approach is known as RNA interference (RNAi). For many years, scientists have been trying to modify inherent activities of plants to improve their yield and quality. Suppression of some biological activities by RNAi would help to achieve such goals.

Key words: Small RNA (sRNA), siRNA, miRNA, posttranscriptional gene silencing.

1. Introduction

In the last decade, the genome sequences of some plants such as *Arabidopsis*, rice, poplar, papaya, soybean, and tomato have been made available. At the same time, we began to understand the function of small RNAs (sRNAs) of 19–25 nt in fine-tuning gene regulation in organisms. The system in which sRNAs suppress expression of specific genes based on sequence specificities is known as RNA interference (RNAi). sRNAs do not occur by chance but are actively produced for normal developmental progression or stress resistance of the organism. Now that we can decipher a rough map of RNAi mechanisms at the molecular level, we are at the stage where we can utilize this mechanism to regulate or modify some biological activities to improve plants or crops. Many different approaches are discussed in this

H. Kodama, A. Komamine (eds.), *RNAi and Plant Gene Function Analysis*, Methods in Molecular Biology 744,
DOI 10.1007/978-1-61779-123-9_1, © Springer Science+Business Media, LLC 2011

book. In this chapter, I briefly overview the historical flow of research on RNAi over the last decade.

2. Discovery of RNA Silencing

2.1. Discovery of PTGS Events

During the long history of the plant sciences, many researchers have attempted to enhance the biological activities of plants. Soon after the establishment of the *Agrobacterium*-mediated transformation system to introduce foreign DNA, a cDNA copy of the coat protein (*CP*) gene of tobacco mosaic virus (TMV) was introduced into tobacco plants. Since wild-type plants do not have the *CP* gene, the transformed plants were forced to accommodate a novel foreign gene from outside their original genome. When the transformants were attacked by TMV, they showed less severe or delayed symptoms compared with the wild-type plants (1). In other words, the transformed plants showed some resistance against TMV and were described as immune to TMV infection. After screening populations of transformants, researchers identified good breeding resources that conferred satisfactory resistance upon crops, even those growing in field conditions.

After this finding, several analogous approaches were applied in various host–virus systems. If a severe crop loss occurred due to infection by a certain virus, researchers determined the cause of the disease, identified the viral pathogen, made a cDNA of the virus, and determined its genome sequence. Based on available databases of virus information, they then assigned coding regions of the CP, replication protein, or movement protein in the genome. The CP coding region was then cloned and placed just downstream the cauliflower mosaic virus 35S promoter in a binary vector, and the crop plant of interest was transformed with the vector. Such transgenic plants showed less severe or delayed symptoms, and this newly acquired resistance could be expected to last for a year or two. This approach was very attractive, since conventional breeding techniques are time consuming and laborious to attain such traits to a similar level. The CP-mediated resistance (CPMR) approach was used to complement conventional breeding approaches, such as searching for gene resources from wild species to improve the germplasm (2). However, although it was generally applicable to many crops and it showed specificity against the original virus, its fundamental mechanism of action remained unclear for some time.

Initially, researchers expected that plants that accumulated higher levels of CP would have stronger immunity to the virus. However, this was not the case. Some plants that accumulated lower levels of CP showed greater immunity than those with higher levels of CP. Later, it was reported that plants expressing other viral gene products, such as the replicase protein, also showed resistance against the original virus. Interestingly, introduction of transgenes in which the original translation initiation codons were modified, or those that were truncated, resulted in virus resistances similar to those observed after the introduction of wild-type virus genes. Whether or not the transgenes were mutated in some way, the resistance of transformed plants was still clearly virus specific.

In a separate line of research, many plant researchers attempted to overproduce proteins or enzymes by introducing multiple copies of the gene of interest into the plant genome of interest. However, contrary to expectations, expression of both endogenous and introduced exogenous genes was decreased, rather than increased, as a result. One typical and classical example is the introduction of the chalcone synthase (*CHS*) gene, a key enzyme in the pigment biosynthetic pathway, into the petunia genome. Transformed petunia plants often showed flowers with paler colours or variegated petals (3). Such results must have caused arguments about whether some technical or careless mistakes were made during the experiment. However, subsequent analyses showed markedly lower levels of transcripts of both the endogenous and the introduced genes in the transformants. A run-off transcription assay using isolated nuclei revealed that their original transcription activity was the same as that in wild-type plants. Thus, transcription from both the endogenous and introduced genes was normal, but transcripts were degraded by some as-yet-unknown events after mRNAs were transcribed. The sequence-specific degradation of mRNA resulted in lower levels of protein expression in transformants than in wild-type plants. The term posttranscriptional gene silencing (PTGS) was coined to describe this phenomenon.

2.2. Discovery of Small RNAs in Plants

It is quite possible that plants recognize the introduction of a foreign gene fragment as an invasion of genetic materials in a similar way that they sense the invasion of pathogens or the activation of parasites. How can a plant 'sense' the introduced gene copy as invading genetic information? How do we explain the sequence specificity? The signal must trigger the following molecular events to attack the invading entity bearing similar genome information.

Green fluorescent protein (GFP) is a useful tool for the analysis of the PTGS phenomenon. Initially, a *GFP* gene was placed under the control of a constitutive 35S promoter to express the protein in all tissues (GFP+plant). When a binary vector for the expression of GFP was locally inoculated into GFP+plants with the aid of *Agrobacterium*, it resulted in the disappearance of the GFP fluorescence. A time-course analysis revealed that the GFP fluorescence disappeared (dark tissues on GFP+ fluorescent tissues) as if a sequence-specific substance spread from the initial infection site to systemic tissues like a 'dark virus'.

In 1999, Hamilton and Baulcombe (4) reported that sRNA moieties 18–25 nt long were found in PTGS-state plants, but not in wild-type plants. In the previous PTGS example in GFP+plants, such sRNAs were detected only when a GFP coding fragment was used as a hybridization probe. This indicated that an antisense sRNA complementary to the GFP sequence was associated with the observed PTGS. This finding could explain the sequence specificity of the PTGS phenomenon. sRNAs with sufficient sequence identity would somehow convey the sequence specificity systemically, like yelling 'Here comes the GFP sequence!'

In addition, in the GFP-PTGS event described above, the abundance of such sRNA molecules increased over several days in GFP+plants that were infected with a GFP-expressing binary vector. Virus resistance PTGS could be now regarded as a type of PTGS.

Subsequently, researchers found that in plants infected with potato virus X (PVX), an sRNA with an antisense sequence appeared after infection in both systemic and inoculated leaves. This led researchers to hypothesize that when plants are infected by a virus, they allow viral replication to some extent but initiate a resistance reaction against its propagation by synthesizing such sRNAs with sequences complementary to the pathogen genome (4).

When sRNA molecules were produced in these cases, it was assumed that they bind to complementary sequences, leading to degradation at the sites where annealing occurred. The detailed mechanism was revealed later by characterization of mutants with lower PTGS. Such sRNAs were now known as short interfering RNA (siRNA).

Even very short RNA molecules, for example, a 21-nt oligoribonucleotide, can distinguish one specific sequence out of 4^{21} (roughly 4×10^{12}) different sequences. This number is greater than the amount of genetic information in one organism. Therefore, the sRNA of a specific sequence can target one specific locus in an entire genome.

Until this discovery, RNA molecules 18–25 nt in length may have been discarded during RNA purification or neglected as degradation products. A thorough understanding of the activities and roles of these molecules could be achieved by studying their different, but fundamentally common, biological effects.

2.3. Discovery of MicroRNAs

RNA molecules known as microRNAs (miRNAs) were first found in a worm, *Caenorhabditis elegans*. A mutant worm, *lin-4* (lineage defective 4), which had a defect in the juvenile-to-adult developmental transition, was characterized and the gene responsible for the defect was isolated. The genome fragment that complemented the mutant phenotype did not encode a protein sequence but was transcribed into RNA in wild-type worms (5). A similar finding followed in another worm mutant, *let-7* (6). In both cases, the first transcripts were processed into smaller RNA molecules and finally into a 21-nt sRNA, which is now known to be an miRNA. miRNAs target mRNA molecules with complementary sequences and inhibit their translation. Lin-4 or let-7 miRNAs negatively regulate protein expression by annealing with corresponding mRNAs (6), resulting in temporal regulation of developmental phase change.

After these two miRNAs were identified, researchers found many 21-nt sRNAs that were miRNA candidates via concurrent analyses of genome sequences of many eukaryotic organisms. Such sRNAs and their roles in biological regulation have been studied in insects, mammals, *Arabidopsis* and rice. The first example showing the biological importance of miRNA in plants was demonstrated by Carrington and Weigel in 2003 (7). They isolated *Arabidopsis* mutants that bear serrated leaves, known as *JAW* mutants, and showed that the gene locus responsible for the mutation did not encode any protein. The candidate transcript had the potential to produce an sRNA moiety, which was subsequently revealed to be an miRNA. The miRNA was produced in the wild-type plants but not in *jaw* mutants (7). The miRNA had a complementary sequence to a part of the mRNA sequence encoding so-called TCP proteins, which are a class of transcriptional regulator proteins. Misregulation of TCP expression resulted in abnormal formation of leaf edges (7).

Several criteria were established to define whether a given sRNA is an miRNA; the sRNA sequences must be encoded in the respective genome, they must be transcribed from large precursor molecules and the precursor molecule should be processed to form the mature molecule. These miRNAs target and base pair with so-called target mRNA sequences (8). It is now widely accepted that specific base pairing between an miRNA and its target gene sequence specifically represses translation of that gene or cleavage of target mRNAs.

3. RNAi Components

siRNAs and miRNAs have similar nucleotide lengths and almost identical chemical properties. Thus, it is no coincidence that the molecular machinery by which siRNA or miRNA is processed or involved in repression of target RNAs is shared between the two RNA molecules. Both sRNAs can anneal or hybridize with target mRNAs and repress their translation. For this reason, these two repression phenomena are collectively referred to as RNA interference (RNAi).

Genetic approaches were applied to isolate and characterize *Arabidopsis* mutants showing impaired PTGS (9). A transgenic *Arabidopsis* line showing PTGS of a 35S-β-glucuronidase transgene was mutagenized using ethyl methanesulfonate. Release from PTGS resulted in approximately 10- to 100-fold increased β-glucuronidase activity. They isolated several suppressor mutations of gene silencing (*sgs1–sgs18*) using this system. Later, it was revealed that many of the genes responsible for *sgs* mutations were also involved in miRNA processing/regulation machinery as well as in PTGS.

Proteins with roles in miRNA regulation were identified by characterization of developmentally defective mutants. Some embryonic-lethal mutants were found to be disruptants of some RNAi components such as so-called *AGO1* or *DCL1* gene products. In addition, some mutants showing different flowers or defects in leaf morphologies were named *hen1*, *sgs3*, *rdr6*, or *dcl4*. Such gene products are involved in maturation and functioning of miRNA or some variant. Changes in the levels of miRNA accumulation resulted in changes in the repression level of target mRNA, leading to abnormalities in temporal development or disruptions of the borders between adjacent organs.

As a general introduction to RNAi, in the following part I will briefly summarize three representative activities that have roles in sequence-specific gene regulation as RNAi components with the aid of various sRNAs (**Fig. 1.1**).

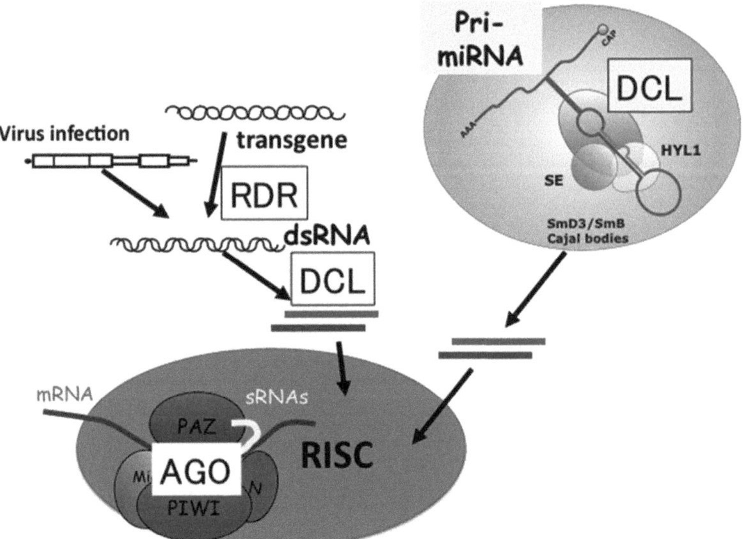

Fig. 1.1. Overview of RNAi inside organisms. Some RNAi events and relevant RNAi components are shown schematically. Abbreviations are described in the text. Intranuclear transcriptional gene silencing is not depicted here.

3.1. RNA-Dependent RNA Polymerase

In PTGS, it is expected that the first trigger RNA sequences must be amplified or transformed into a double-stranded RNA form. RNA-dependent RNA polymerases (RDRs) are plant-specific host-encoded enzymes that synthesize antisense RNA sequences using single-stranded RNAs as templates (10). Such double-stranded RNAs are not the normal RNA moiety in cells. In fact, they stimulate 'amplifier' molecules to invoke the RNAi 'defence' system. There are six members of RDRs in *Arabidopsis.* It is assumed that PTGS requires the activities of RDRs as an amplifier. In *Arabidopsis*, RDR6 is the main enzyme involved in the onset of PTGS (10). In contrast, miRNA-mediated repression does not require RDR activity. The precursor miRNA molecule are transcribed and produced from genome and forms an intramolecular stem–loop structure with a partial double-stranded character by itself.

3.2. Dicer-Like Proteins and Double-Stranded RNA-Binding Proteins

Double-stranded RNA would be processed, whether fully or partially, into the 21–24-nt sRNAs, siRNAs, or miRNAs by so-called Dicer-like proteins (DCLs) (10). These proteins have RNase III, PAZ, and double-stranded RNA-binding (dsRBD) domains, and a very large molecular mass of not less than 100 kDa. There are four members of DCLs in *Arabidopsis.* In PTGS, a fully or partial double-stranded RNA intermediate would be processed into an siRNA duplexes by one of the DCLs, DCL4 (10). In contrast, another DCL, DCL1, is mainly involved in miRNA maturation to form miRNA/miRNA*, which has two base overhangs at both 3′-ends and a phosphate at each 5′-end (11). DCL1 must interact with a partner molecule containing two dsRBD domains, known as hyponastic leaves 1 (HYL1), to precisely cleave precursor molecules in succession. In *dcl1* or *hyl1* mutant backgrounds, the miRNA precursor molecules are transcribed at normal levels but are poorly or imprecisely processed into miRNAs (12). Analyses of precursor, intermediate, and mature miRNAs in different genetic backgrounds revealed that precise cleavage relies on the interaction between DCL1 and HYL1 (12).

3.3. ARGONAUTE Proteins

sRNA duplexes are first incorporated into the so-called RNA-induced silencing complex (RISC). In plants, the exact composition of the RISC is still unknown, but the main component is an ARGONAUTE protein (AGO). There are 10 members of AGOs in *Arabidopsis.* One strand of the duplex is entrapped by an AGO protein via thermodynamic stability or the 5′ base preference of the AGO (13, 14). Then, translational repression or endonucleolytic cleavage of the duplex occurs, based on its sequence information. It was experimentally shown that AGO1 and 4 have slicing activities in these complexes (15), but not clear whether other

AGOs have similar activities. Slicer activity cleaves target RNAs based on complementarity with antisense sRNA sequences.

4. Application of RNAi Phenomena

When the molecular mechanisms of sRNA are revealed, we realize that RNAi phenomena or events are already applied empirically in many circumstances. As described in the first section, CPMR is partly or fully mediated by an RNAi mechanism in which the introduced coat protein gene stimulates the RNAi machinery, preparing it for the virus challenge. Even before transgenic approaches were used, farmers in some areas protected their crops by preinoculation with mild attenuated virus strains before attack of more devastating severe strains. Preinoculation with an attenuated strain would trigger a state in which the RNAi system would resist invasion of a virulent virus with a similar sequence. Such approaches are still applied in the cultivation of many cucurbit and solanaceous vegetables and citrus fruit to reduce decreases in yield and quality due to virus infection. The genomes of most plant viruses are single-stranded RNAs with plus polarity. At the initial stage of infection, antisense full-length RNAs are synthesized using the genomic RNA as a template, and subsequently, some full-length, double-stranded RNA molecules are formed. Such double-stranded RNAs are recognized by some DCL and then cleaved into 21–24-nt sRNAs (16). These sRNAs are then entrapped by ARGONAUTE proteins and possibly induce or promote slicer activity, which damages the genome of the challenging virus.

If a foreign gene sequence was inserted into a virus genome, some sRNAs with antisense sequence information of the foreign gene would be formed, since plants cannot discriminate the origins of the sequences. Accordingly, the plants would then repress expression of the foreign gene as well as virus multiplication. If the foreign gene originated from the plant species itself, the recombinant virus infection would result in the repression of the corresponding endogenous gene (17). The phenomenon is known as virus-induced gene silencing. This technique could be used as an easy and rapid method to knock down the function of a particular gene, because the time-consuming establishment of transgenic plants would not be necessary.

As in the case of *CHS*–petunia described earlier, overexpression of a sense-strand sequence suppressed expression of the corresponding gene. This approach, known as sense-PTGS (S-PTGS), is not as complicated as other methods. However, while introduction and expression of a target sequence in the sense orientation suppresses gene expression to some extent, it is not as reliable or efficient as some other methods. Instead, some researchers introduced the target sequences in tandem but in opposite directions with an intervening sequence as a spacer to make an intramolecular stem–loop structure. It was anticipated that such RNA structures would be more susceptible to DCL activities. This approach resulted in more consistent PTGS effects (18). This is possibly because the duplex would remove the need for RDR involvement, because it would be degraded directly by some DCL proteins and invoke the RNAi system. This system, which is known as inverted-repeat posttranscriptional gene silencing (IR-PTGS), is used frequently as described in other chapters.

Understanding the process of miRNA maturation led to construction of an artificial miRNA to suppress expression of desired sequences. Precursors of endogenous miRNA were modified by site-directed mutagenesis to exchange the miRNA/miRNA* regions with the desired sequences to obtain artificial microRNAs (amiRNAs) (19). In the original example, leafy mRNA was targeted with an amiRNA constructed in the miR172 backbone, and this achieved the desired results. The amiRNA approach was shown to be applicable in many cases.

Trans-acting small interfering RNAs (tasiRNAs) are a class of siRNAs, but their biological functions are similar to those of miRNAs. They can hybridize with target mRNAs through their own 21-nt sequences and suppress the expression of target sequences. Production of tasiRNAs is initiated by miRNA-assisted specific cleavage of TAS RNAs, which are transcribed from noncoding regions of the genome (20). The cleaved products are subsequently guided by SGS3 and converted into double-stranded RNAs via the action of RDR6. The dsRNA is then processed into 21-nt successive siRNAs in the same phase by DCL4 (16). The process is somewhat complex, but the tasiRNA system can be modified in a more flexible way compared to the miRNA system to construct artificial tasiRNA to suppress target sequences.

5. Conclusion and Future Perspectives

Until recently, RNAi has been used empirically by expressing target sequences in various ways putting PTGS in consideration. In some examples, artificial double-stranded RNAs were forced to be expressed in plants, then leading to suppression of homologous sequences in chromatin by DNA or histone modification, a phenomenon called transcriptional gene silencing (TGS). siRNAs are made from dsRNAs through the action of DCL3 and they would repress the local chromatin activity employing an AGO member, AGO4, in this RNA-directed DNA methylation (21). Recently such phenomenon is also included in RNAi as well and its application is now being challenged. Together, we are getting the picture of the basic components that are involved in a variety of RNAi's (both PTGS and TGS) in plants. This brings us to a new stage where we can formulate more elegant ways to upregulate or downregulate specific plant genes in organisms. Researchers are now able to manipulate the model precursor molecule to fit each application more reliably. Readers will be able to overview various trials and specific issues in the following chapters of this book.

Acknowledgements

I would like to thank many collaborators and colleagues who worked in Watanabe lab over the years. As well, I thank the Ohio Arabidopsis Seed Stock Center for providing valuable seeds for our research.

References

1. Powell, P. A., Nelson, R. S., De, B., Hoffmann, N., Rogers, S. G., Fraley, R. T., and Beachy, R. N. (1986) Delay of disease development in transgenic plants that express the tobacco mosaic virus coat protein gene. *Science* **232**, 738–743.
2. Truniger, V. and Aranda, M. A. (2009) Recessive resistance to plant viruses. *Adv. Virus Res.* **75**, 119–159.
3. Jorgensen, R. (1990) Altered gene expression in plants due to trans interactions between homologous genes. *Trends Biotechnol.* **8**, 340–344.
4. Hamilton, A. J. and Baulcombe, D. C. (1999) A species of small antisense RNA in posttranscriptional gene silencing in plants. *Science* **286**, 950–952.
5. Lee, R. C., Feinbaum, R. L., and Ambros, V. (1993) The *C. elegans* heterochronic gene *lin-4* encodes small RNAs with antisense complementarity to *lin-14*. *Cell* **75**, 843–854.
6. Moss, E. G. (2000) Non-coding RNA's: Lightning strikes twice. *Curr. Biol.* **10**, R436–R439.
7. Palatnik, J. F., Allen, E., Wu, X., Schommer, C., Schwab, R., Carrington, J. C., and Weigel, D. (2003) Control of leaf morphogenesis by microRNAs. *Nature* **425**, 257–263.
8. Meyers, B. C., Axtell, M. J., Bartel, B, Bartel, D. P., Baulcombe, D., Bowman, J. L., Cao, X., Carrington, J. C., Chen, X., Green, P. J., Griffiths-Jones, S., Jacobsen, S. E., Mallory, A. C., Martienssen, R. A., Poethig, R. S., Qi, Y., Vaucheret, H., Voinnet, O., Watanabe, Y., Weigel, D., and Zhu, J.-K. (2008) Criteria for annotation of plant microRNAs. *Plant Cell* **20**, 3186–3190.
9. Elmayan, T., Balzergue, S., Béon, F., Bourdon, V., Daubremet, J., Guénet, Y., Mourrain, P., Palauqui, J.-C., Vernhettes, S., Vialle, T., Wostrikoff, K., and Vaucheret, H. (1998) *Arabidopsis* mutants impaired in cosuppression. *Plant Cell* **10**, 1747–1757.
10. Dunoyer, P., Himber, C., and Voinnet, O. (2005) DICER-LIKE 4 is required for RNA interference and produces the 21-nucleotide small interfering RNA component of the plant cell-to-cell silencing signal. *Nat. Genet.* **37**, 1356–1360.
11. Xie, Z., Johansen, L. K., Gustafson, A. M., Kasschau, K. D., Lellis, A. D., Zilberman, D.,

Jacobsen, S. E., and Carrington, J. C. (2004) Genetic and functional diversification of small RNA pathways in plants. *PLoS Biol.* **2**, E104.

12. Kurihara, Y., Yuasa, T., and Watanabe, Y. (2006) The interaction between DCL1 and HYL1 is important for efficient and precise processing of pri-miRNA in plant microRNA biogenesis. *RNA* **12**, 206–212.
13. Mi, S., Cai, T., Hu, Y., Chen, Y., Hodges, E., Ni, F., Wu, L., Li, S., Zhou, H., Long, C., Chen, S., Hannon, G. J., and Qi, Y. (2008) Sorting of small RNAs into *Arabidopsis* Argonaute complexes is directed by the 5′ terminal nucleotide. *Cell* **133**, 116–127.
14. Takeda, A., Iwasaki, S., Watanabe, T., Utsumi, M., and Watanabe, Y. (2008) The mechanism selecting the guide strand from small RNA duplexes is different among Argonaute proteins. *Plant Cell Physiol.* **49**, 493–500.
15. Baumberger, N. and Baulcombe, D. C. (2005) *Arabidopsis* ARGONAUTE1 is an RNA slicer that selectively recruits microRNAs and short interfering RNAs. *Proc. Natl. Acad. Sci. USA* **102**, 11928–11933.
16. Gasciolli, V., Mallory, A. C., Bartel, D. P., and Vaucheret, H. (2005) Partially redundant functions of *Arabidopsis* DICER-like enzymes and a role for DCL4 in producing *trans*-acting siRNAs. *Curr. Biol.* **15**, 1494–1500.
17. Thomas, C. L., Jones, L., Baulcombe, D. C., and Maule, A. J. (2001) Size constraints for targeting post-transcriptional gene silencing and for RNA-directed methylation in *Nicotiana benthamiana* using a potato virus X vector. *Plant J.* **25**, 417–425.
18. Beclin, C., Boutet, S., Waterhouse, P., and Vaucheret, H. (2002) A branched pathway for transgene-induced RNA silencing in plants. *Curr. Biol.* **12**, 684–688.
19. Schwab, R., Ossowski, S., Riester, M., Warthmann, N., and Weigel, D. (2006) Highly specific gene silencing by artificial microRNAs in *Arabidopsis. Plant Cell* **18**, 1121–1133.
20. Yoshikawa, M., Peragine, A., Park, M. Y., and Poethig, R. S. (2005) A pathway for the biogenesis of *trans*-acting siRNAs in *Arabidopsis. Gene Dev.* **19**, 2164–2175.
21. Matzke, M., Kanno, T., Daxinger, L., Huettel, B., and Matzke, A. J. (2009) RNA-mediated chromatin-based silencing in plants. *Curr. Opin. Cell Biol.* **21**, 367–376.

Chapter 2

Caveat of RNAi in Plants: The Off-Target Effect

Muthappa Senthil-Kumar and Kirankumar S. Mysore

Abstract

RNA interference (RNAi), mediated by short interfering RNAs (siRNAs), is one of the widely used functional genomics method for suppressing the gene expression in plants. Initially, gene silencing by RNAi mechanism was believed to be specific requiring sequence homology between siRNA and target mRNA. However, several recent reports have showed that non-specific effects often referred as off-target gene silencing can occur during RNAi. This unintended gene silencing can lead to false conclusions in RNAi experiments that are aimed to study the functional role of a particular target gene in plants. This especially is a major problem in large-scale RNAi-based screens aiming for gene discovery. Hence, understanding the off-target effects is crucial for minimizing such effects to better conclude gene function analyzed by RNAi. We discuss here potential problems of off-target gene silencing and focus on possibilities that favor this effect during post-transcriptional gene silencing. Suggestions to overcome the off-target effects during RNAi studies are also presented. We believe that information available in present-day plant science literature about specificity of siRNA actions is inadequate. In-depth systematic studies to understand their molecular basis are necessary to enable improved design of more specific RNAi vectors. In the meantime, gene function and phenotype results from present-day RNAi studies need to be interpreted with caution.

Key words: Off-target prediction, plant functional genomics, RNA silencing, siRNA scan, unintended silencing.

1. Introduction

RNA interference (RNAi) technology has made pan-genomic functional gene analysis a reality and it is a powerful strategy for gene discovery and validation (1). RNAi is dominant, so the gene-silenced phenotypes can be observed in the T1 generation itself. RNAi often leads to partial knockdown of gene transcripts, thus providing a range of phenotypes that may differ in severity. RNAi facilitates the study of essential genes whose complete

H. Kodama, A. Komamine (eds.), *RNAi and Plant Gene Function Analysis*, Methods in Molecular Biology 744,
DOI 10.1007/978-1-61779-123-9_2, © Springer Science+Business Media, LLC 2011

inactivation would lead to lethality or extremely severe pleiotropic phenotypes. RNAi can be quickly and easily used in a wide range of genotypes or even species, whereas identification of gene mutations is limited to certain plant species which has the mutant resources. Expression of RNAi constructs can be controlled in a tissue-specific or time-dependent manner (2). There are several examples where loss of gene function results in improved plant performance (3–6). Thus, recently RNAi has been effectively used in transgenic crop development for commercial uses (3, 4, 6). One of the characteristics of RNAi is its ability to silence genes in a sequence-specific manner. However, recent evidences suggest that small interfering RNA (siRNA; RNAi pathway intermediate) does not always target a specific gene, thus resulting in non-specific gene silencing (7, 8). Non-specificity can occur when partial sequence homology allows siRNA to degrade mRNA for genes that are not the intended silencing targets. Several other deviations that can possibly occur at different steps during post-transcriptional gene silencing (PTGS) pathway can also favor such non-specific effect (7, 9–12). Such non-specific effect of siRNA to degrade mRNA for genes that are not the intended silencing targets is referred as off-target silencing.

Recently our group has reported potential possibilities of off-target silencing effects during post transcriptional gene silencing (PTGS) in *Arabidopsis* and *Nicotiana benthamiana* (8). Off-target effects have also been widely observed from large-scale screens in animals (7, 9). RNAi is a rapidly growing field with considerable interest to the plant science community for potential genomics research and agriculture applications. Hence, the potential deleterious off-target silencing needs to be eliminated. In this review, we attempt to provide detailed information about the nature of off-target silencing and also propose ways to overcome this effect.

2. RNAi Mechanism and dsRNA Delivery into Plants

Long double-stranded RNA (dsRNA) can be used to silence the expression of target genes. Upon introduction, the dsRNAs enter a cellular pathway that provokes PTGS. The dsRNAs get processed into 21–23 nucleotide siRNAs by an RNase III-like enzyme called Dicer. In plants, Dicer-like 4 (DCL4) is predominantly involved in this step (13). Then the siRNAs assemble into endoribonuclease-containing complexes known as RNA-induced silencing complex (RISC). The siRNA strands subsequently guide the RISCs to complementary mRNA molecules, where they cleave and destroy the cognate mRNA. Cleavage of cognate mRNA takes place near the middle of the region bound by the siRNA strand. These steps are elaborately reviewed elsewhere (1, 14, 15).

RNAi in plants can be achieved by expressing hairpin RNA (hpRNA) that folds back to create a dsRNA. These hpRNAs are potent inducers of PTGS and give rise to siRNAs derived from the dsRNA. Transforming plants with dsRNA-expressing vectors for the selected gene by *Agrobacterium*-mediated transformation or particle bombardment, infecting plants with viral vectors that can express dsRNA (virus-induced gene silencing), infiltrating *Agrobacterium* cultures harboring hpRNAi construct for transient gene silencing are commonly used methods for gene silencing in plants (8, 16–18). Most of these methods require vector construction and plant transformation. Direct siRNA delivery into plants is another method proposed for large-scale RNAi screens as in animal system. This method has been demonstrated using laser-induced stress wave in plant cell cultures (19). Apart from these, artificial microRNA (amiRNA)-based vectors have been recently shown to be effective for gene silencing (20). These vectors upon delivery into plant cells provoke the PTGS and are expected to silence genes of interest.

3. General Adverse Effects of Off-target Gene Silencing

RNAi has been widely used as a reverse genetics tool for gene function characterization in plants. Our earlier study (8) computationally predicted that about 50–70% of gene transcripts in *Arabidopsis* plants have potential off-targets when used as silencing trigger for PTGS and this could obscure experimental results. Up to 50% of the predicted off-target genes tested in plants were actually silenced when tested experimentally (8). Apart from other studies (9, 11, 20), our group has also demonstrated that such off-target silencing lead to difficulties in identifying exact functional role of target genes by using stable RNAi or VIGS approaches (8).

Another use of RNAi or VIGS is its application in large-scale forward genetic screens for gene discovery and functional analyses. Impact of off-target silencing during large-scale screening in animals has been demonstrated (7, 9). We hypothesize that potential for off-target silencing also exists in high-throughput screens in plants (e.g., AGRIKOLA – *Arabidopsis* genomic RNAi knockout line analysis, http://www.agrikola.org/index.php?o=/agrikola/main or http://www.chromdb.org/) and is not yet reported simply because of lack of systematic study.

Yet another important application of RNAi is in the area of crop improvement. Today, researchers are engineering a variety of crops to produce siRNAs that will silence essential genes in insect pests, nematodes, and pathogens, an approach called host-delivered RNAi (HD-RNAi) (3, 4, 21). Plant genes are known to share at least partial sequence similarity with animal genes. Specific domains of certain genes are conserved across different organisms. If off-targeting can unexpectedly silence genes in plant or pest, such unintended effects will raise concern not only about plant phenotypic pleiotropy but also in the organisms and the subsequent environmental consequences. In theory, vertical gene flow of an RNAi-mediated pollen lethality phenotype to native plants could alter fitness and biodiversity (3) and off-target silencing will only boost such negative effects. Hence this could be a potential biosafety issue for RNAi transgenic plants developed for crop improvement. Off-target silencing also poses threat to appli-

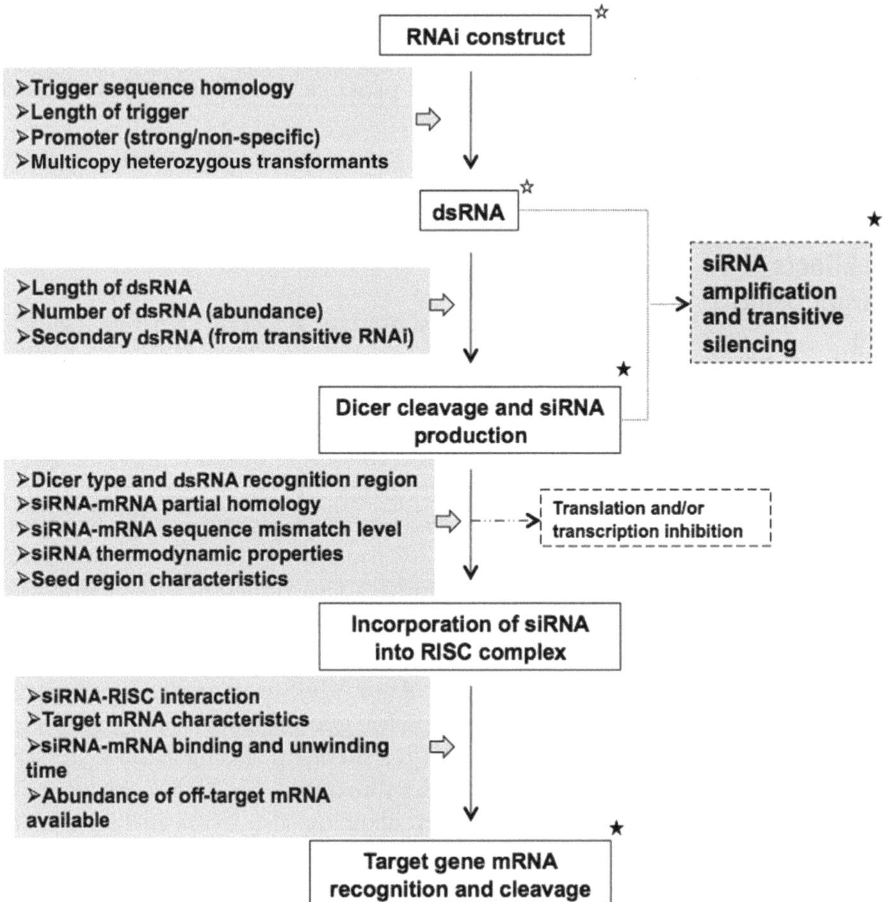

Fig. 2.1. Possible sources of off-target effects during RNAi in plants. The flow diagram shows RNA interference (RNAi) in transgenic plants with hpRNA vector along with potential possibilities of off-target induction in each step. ☆ Steps that can be manipulated to increase specificity of silencing. ★ Steps that cannot be manipulated at this time due to lack of information regarding mechanism.

cation of RNAi in developing several food crops with improved nutritional value. Especially, off-target effects if not controlled can severely compromise biosafety in RNAi plant food consumers, namely human and livestock. We discuss below the steps of PTGS pathway in plants that could potentially contribute to off-target gene silencing (*see* **Fig. 2.1**).

4. Steps in PTGS and Their Relation to Off-target Activity

4.1. RNAi Vector, Silencing Trigger, and dsRNA

To develop genetically stable RNAi plants, binary vectors (called hpRNAi vectors) with gene fragments cloned in invert orientation separated by a short intron are widely used (8, 18, 22). The gene fragment used for producing dsRNA to silence the intended target mRNA is called as the "trigger." Trigger sequence in RNAi vector plays important role in deciding specificity of gene silencing. Normally 250–350-bp fragment of gene is optimum as a trigger for hpRNAi vectors, while smaller lengths are also used in other methods, namely direct siRNA delivery and amiRNA vectors (8, 17–20, 22). First, length of dsRNA correlates with both silencing efficiency (23) and off-target silencing (11). While long hairpins are more likely to generate a diverse set of optimally effective siRNAs, they also have an increased potential to produce siRNAs with off-target effects (24). Essentially the chance of off-target increase with greater length of the initial dsRNA sequence has been demonstrated (11, 20). Second, number of dsRNA also has correlation with off-target silencing. Higher dsRNA than required threshold is known to favor off-target effects (7, 22, 25). Sometimes RNA-dependent RNA polymerase (RDR)-mediated amplification using small RNA or piece of dsRNA produces secondary dsRNA (26, 27). This can considerably influence the dsRNA number in cell. Apart from this, in the genetically stable RNAi plants, position of RNAi construct integration in the plant genome and copy number are factors deciding extent of its transcript expression (28, 29), thereby influencing the number of dsRNA production. To a certain extent, copy number and position of RNAi vector are influenced by the method used for plant transformation for delivering RNAi vectors. Third, use of strong promoters in RNAi vectors to drive hpRNA production could favor a completely different type of off-target effect owing to the inhibition of natural miRNA or siRNA regulation through saturation of the pathways with exogenous or transgene siRNAs (30). Fourth and most important aspect influencing off-target silencing is the nature of trigger sequence used in RNAi vector. Off-target silencing is mainly influenced by trigger sequence homology with mRNA. We have shown that

21-nucleotide stretch of 100% identity between the silencing trigger sequence and endogenous target gene sequences is not absolutely required to provoke gene silencing (8, 31). Computational tools also predicted the influence of trigger sequence used in hpRNAi or miRNA vector construction or siRNA design in off-target gene silencing. Apart from these, the silencing trigger sequence in hpRNA vector selected from the 3′ UTR region of a gene is known to favor off-target gene silencing compared to a trigger from the 5′ region (11).

4.2. Dicer Cleavage and siRNA Production

In RNAi *Arabidopsis* plants, dsRNA is predominantly processed by DCL4 to produce specific siRNA. However, other dicers, namely DCL2 and DCL3, can also be involved at this step and produce less specific siRNA, leading to unpredictable gene silencing (30). DCL action eventually decides length of siRNA, its diversity, and quantity (10). The dsRNA length and overhang features also influence dicer cleavage. For example, to efficiently process a short dsRNA, certain features like short overhanging 3′ nucleotides on the dsRNA are beneficial and increase specificity. Further, excess unprocessed short dsRNAs can bind to mRNA without cleaving the molecule, preventing translation of the mRNA to protein. Such short dsRNAs can also bind to DNA, inhibiting the transcription of DNA to mRNA (24). The molecular mechanism behind these two actions is not yet completely understood and not reported in plants.

4.3. siRNA Characteristics and Transitive Silencing

siRNAs are the ultimate determinants of specificity of silencing. First and foremost, difficulty in maintaining targeted gene silencing is the limited sequence specificity of siRNAs. As few as 14 nucleotides (or even less) of sequence complementarity between siRNA and mRNA can lead to the inhibition of gene expression (8, 9, 11, 31, 32). Limited sequence specificity can potentially favor chances of siRNA sharing homology with off-target mRNA (30). A second specificity problem can occur via "transitive silencing," whereby RNAi against a gene-specific sequence spreads into neighboring sequences (26). In this process, the primary siRNAs bind to complementary transcripts and generate new dsRNAs via a 5′ to 3′ extension carried out by RDR. These newly synthesized dsRNAs can be processed into less specific secondary siRNAs, leading to silencing of any complementary mRNA. This process is called transitive RNAi and can affect the specificity of silencing. Transitive RNAi has been documented in many studies as a cause of non-specific silencing (23, 26, 27, 33). Further, this process can essentially amplify siRNAs, leading to off-target silencing. Such silencing effects can in some cases be transmitted from cell to cell and even over long distances throughout the plant, leading to a potential loss in control over the intended silencing signal (26, 30). An already non-specific source of siRNA or dsRNA under these circumstances can further increase off-target effects and complicate understanding of target gene function from such an RNAi study. Hence, it is important to titrate the level of siRNA to lowest effective level to considerably reduce off-target silencing (7).

Apart from influencing PTGS, the off-target effects at siRNA level can also occur by other mechanisms. Excessive siRNA may function as miRNA and suppress translation without affecting transcript levels (12). Further, any siRNAs with only partial homology match with target mRNA can only loosely bind, leading to hindrance for ribosome access. This can also lead to prevention of protein synthesis by inhibiting translation without RNA silencing (34). Two other types of off-target effects include siRNA cross-hybridizing with undesired mRNA causing degradation and the binding of siRNA to protein (like transcription factors or other transcriptional regulatory enzymes) as an aptamer affecting transcription (34).

In addition, silencing potency of a perfectly matched siRNA is largely determined by the thermodynamic property of its primary sequence (20, 35, 36). Physical and chemical properties of siRNA are important to maintain specificity in gene silencing. GC content and stability of siRNA also play role in silencing and off-target activity (10). Off-target transcript silencing is widespread and mediated largely by limited target sequence complementarity to the seed region of the siRNA guide strand (32, 36, 37). A distinct hexamer sequence of the siRNA guide strand decides the pairing with complementary sequences in the off-targeted mRNA. This hexamer region between 2 and 7 nucleotides is called seed region. The siRNAs that has seed regions with high complement frequency with off-target mRNA can increase chances of off-target silencing. Modifications within siRNA seed region have been shown to reduce both the number of off-target transcripts (by 66%) and the magnitude of their regulation, without significantly affecting silencing of the intended targets (32). Seed region influence on off-target silencing is not widely reported in plants but has the potential to occur. When this phenomenon was investigated in more detail in mammalian cells, a strong correlation between off-targets and short stretches of complementarity to the siRNA in the 3′ UTRs of the affected transcripts was found (37). Certain position-specific, sequence-independent chemical modifications in siRNA has also been shown to reduce off-target effects in animals (32).

4.4. RISC Complex and mRNA Target Search Activity

Generally, the antisense siRNA (guide strand) directs RISC to complementary mRNA, while the second (sense) strand is degraded. Off-target effects can be caused by insertion of the sense siRNA strand into the RISC complex instead of the antisense strand (7, 9, 35). During RNAi the active RISC complex with siRNA samples target sites in mRNA non-specifically until the correct one is found. Once the correct target RNA is found, conformational changes occur that enhance RISC catalytic efficiency, leading to slicer activity (32, 35). However, binding of imperfectly matched targets can contribute to other forms of off-target effects, namely inhibition of translation (no slicing). Modifications to siRNAs that weaken or disrupt RISC–mRNA interaction in the seed region have been shown to reduce off-target silencing in human (32). Within the short sequence of siRNA, six base pairs (seed) play an important role in the binding of target mRNA to RISC. Studies by Jackson and colleagues have also shown that this seed region of siRNA shows the highest correlation with off-target effects (32).

4.5. Target Gene mRNA Characteristics and Sequence Homology Required for Silencing

Apart from siRNA–mRNA homology requirements, off-target silencing also depends on characteristics of target mRNA. This includes sequence context surrounding complementary regions, position of the complementary region in mRNA, and copy number of the complementary region (10). Studies suggest that each target sequence possesses an inherent degree of susceptibility to dsRNA-mediated degradation (28). Highly expressed genes are more amenable for unregulated silencing (28) as the frequency of activated RISC accessing their mRNA will be high. Yet another aspect that compromises the siRNA–mRNA specificity is nucleotide base pair mismatch, although plants can tolerate small number of mismatches compared to animals (10, 11, 24, 31). According to Warthmann and colleagues (20), any transcript other than the specified target, which matches the 21-mer with five mismatches, is considered an off-target (20). The efficiency of off-targeting due to mismatch depends on the length of mismatching nucleotides and their position on siRNA (11). Although only few of above-described steps in PTGS can be controlled by researchers to increase specificity of silencing during RNAi, future research can potentially facilitate control of other steps.

5. Confirming the Specificity of Target Gene Silencing in RNAi Experiments

Present RNAi studies in plants are likely to have unintended gene silencing effect (3, 8, 20). Hence ascertaining the specificity of silencing at the end of each RNAi experiment before interpreting results is important. Following are some useful tips: (a) Confirm reduction of target mRNA and its protein expression (without affecting related gene expression); (b) confirm specificity of the observed phenotype; (c) perform gene expression profiling (e.g., microarray) to assay the expression pattern of non-target genes. Jackson and colleagues (7) have shown that expression profiling in conjunction with gene silencing by RNAi will provide an effective means to identify and characterize specific target gene function. (d) Rescue the RNAi effect by expressing an siRNA-resistant copy of target gene. Since most codons in the mRNA of the target gene can be altered at the third nucleotide position without changing the encoded amino acid, the RNAi-resistant copy of native mRNA can be designed to escape silencing (38). This should complement RNAi-induced effects and revert back to wild-type like plant. (e) Confirm that the same phenotype or change in specific metabolic process is produced when other regions of same gene are used as silencing trigger. (f) Examine multiple independent RNAi plants to check for a reproducible phenotype. (g) Compare the results with plant transformed with RNAi vector harboring non-plant gene (e.g., *GFP*). Monitor changes due to plant immune response (while using VIGS). (h) When silencing a gene family member, make sure that other family members are not silenced and vice-versa.

While performing these confirmatory experiments to ascertain the specificity of gene silencing, certain inherent limitations described below need to be kept in mind. (i) Certain genes are known to be induced upon perception of RNA silencing. Transcript reduction is not always directly correlated with phenotype or protein reduction. (ii) Silencing of certain genes may not show a visible phenotype. (iii) The full extent of the contribution of off-target gene regulation to phenotypic induction is not known. (iv) Similar to miRNAs, the effect of siRNAs on off-target protein regulation might be even greater than the effect on off-target transcript silencing. (v) Small changes in the expression levels of some proteins, such as transcription factors, might translate to large effects on phenotype. Until proteomic and metabolomic analyses of siRNA experiments are performed, we will not know the full consequences of siRNA off-target activity. (iv) Many endogenous genes are regulated by natural RNAi pathways, so perturbations in these processes can lead to pleiotropic effects (30).

6. Remedy for Off-target Silencing

(a) Choose highly specific trigger sequence for RNAi vector. This can be done by considering several siRNA characteristics apart from sequence similarity alone. Minimize homology of trigger sequence in RNAi construct to non-target mRNA. If off-targets are predicted using computational approach, select the trigger sequence which can avoid perfect siRNA stretches of less than 11 bp for predicted off-target gene. This should prevent off-target effects while maintaining adequate complementarity for target gene sequence. Computational off-target prediction tools are described in **Section 7**. (b) Wherever possible, use RNAi vectors with tissue-specific and inducible promoters to minimize off-target silencing. Although silencing might spread to other tissues, the benefit of specific promoters to increase specificity of gene silencing has been demonstrated (2, 39). (c) Maintain lowest effective number of dsRNA and siRNA. This may be achieved by having appropriate promoters for RNAi plant generation and by analyzing single-copy homozygous transgenic plants. (d) Analyze many independent biological replicates and account for possible variations. (e) Since structurally different, RNAi-resistant, mRNA-producing construct that has same function with native gene can be artificially synthesized, certain off-target gene effects can be complemented by stable transgene development with RNAi-resistant, off-target gene construct. This method can be applied in commercially important RNAi transgenic plants to nullify off-target effect that occurred due to one or two genes.

Apart from these, some of following aspects could be considered when the adequate information and technology will become available in the future to control off-target silencing. (f) Minimize secondary siRNA production and transitive gene silencing. (g) Titrate the dsRNA and siRNA population to optimum numbers. (h) Minimize incorporation of the siRNA sense strand into RISC.

7. Predicting Off-target Gene Silencing Using Computational Methods

Computational prediction tools basically use sequence similarity to identify effective trigger sequence in RNAi vector. They are less expensive to implement and permit the extension of real parameters into wider ranges for fully observing the trends and effects upon RNAi specificity (11). BLAST has been the most commonly used tool to study homology between sequences. This can be utilized to search for sequences similar to the proposed siRNA in the desired organism. However, because of the short length of siRNAs, using BLAST requires attention to details and potentially relevant alignments can be easily missed. Hence, siRNA design rules were devised to check for sequence similarity with other genes, as well as to prevent the use of sequences that repeat or contain common binding sites, and look at the effect of the positioning of the siRNA on the target sequence. These design rules are usually incorporated into algorithms to search for siRNAs that have matches for all gene sequences in addition to the targeted gene (8, 11).

Our group has developed a publicly available Web-based computational tool called siRNA Scan (http://bioinfo2.noble.org/RNAiScan.htm) to identify potential off-targets during PTGS in plants (8). The input parameters of the tool for analysis of siRNAs and off-targets can be adjusted by users. This siRNA scan tool is useful to design better constructs for PTGS by minimizing off-target gene silencing in plants. This tool includes the option of searching potential off-targets with complete sequence identity or reverse complementary identity of 18–29 nt to the trigger sequence, as well as allowing a few mismatches to the potential siRNAs. For designing amiRNA, a program called WMD2 (http://wmd2.weigelworld.org) is being used (20). Thermodynamics of the trigger sequence interaction with target mRNA can be analyzed using the program RNAup (36). Given the small degree of similarity implicated in off-target gene regulation, it may be difficult to select an siRNA sequence that will be absolutely specific for the target of interest merely based on sequence homology and few other above-described siRNA characteristics. Hence, better methods should be researched. New methods for creating RNAi vectors without non-specific effects need to look at interactions in the siRNA:RISC complex and how modifications of the siRNA can affect RISC. The computational tools will have full-fledged application once the nucleotide sequence information for vast majority of crop species becomes available.

8. Future Perspectives

Plants adopt silencing strategy via miRNA pathway to control genes involved in organelle development and response against environmental stress through miRNA (13, 24, 35). The endogenous native expression of miRNAs in animals and plants does not create adverse effects but perform the specifically programmed role of these miRNAs in gene regulation. How does miRNAs maintain specificity in silencing? What are the mechanisms that regulate these naturally produced small RNAs? Future plant science research should be aimed to fully understand these questions. This could potentially reveal many ways to manipulate RNAi pathway to minimize off-target silencing. Further, a greater understanding of overall RNAi mechanisms specific to plants will allow for more possibilities in reducing off-target effects. Once the specificity of the siRNA is guaranteed, the objectives to use RNAi in studying gene function and other uses can be comfortably widened.

Acknowledgments

RNAi- and VIGS-related projects in KSM laboratory are supported by the Samuel Roberts Noble Foundation, National Science Foundation (grant no. 0445799), Oklahoma Center for the Advancement of Science and Technology (grant no. PSB09-020), and the US–Israel Binational Agricultural Research Development Fund (BARD grant no. IS-3922-06).

References

1. Waterhouse, P. M. and Helliwell, C. A. (2003) Exploring plant genomes by RNA-induced gene silencing. *Nat. Rev. Genet.* **4**, 29–38.
2. Shuai, C., Daniel, H., Uwe, S., and Frederik, B. (2003) Temporal and spatial control of gene silencing in transgenic plants by inducible expression of double-stranded RNA. *Plant J.* **36**, 731–740.
3. Auer, C. and Frederick, R. (2009) Crop improvement using small RNAs: applications and predictive ecological risk assessments. *Trends Biotechnol.* **27**, 644–651.
4. Gordon, K. H. J. and Waterhouse, P. M. (2007) RNAi for insect-proof plants. *Nat. Biotechnol.* **25**, 1231–1232.
5. Niu, Q.-W., Lin, S.-S., Reyes, J. L., Chen, K.-C., Wu, H.-W., Yeh, S.-D., and Chua, N.-H. (2006) Expression of artificial microRNAs in transgenic *Arabidopsis thaliana* confers virus resistance. *Nat. Biotechnol.* **24**, 1420–1428.
6. Tang, G. and Galili, G. (2004) Using RNAi to improve plant nutritional value: From mechanism to application. *Trends Biotechnol.* **22**, 463–469.
7. Jackson, A. L., Bartz, S. R., Schelter, J., Kobayashi, S. V., Burchard, J., Mao, M., Li, B., Cavet, G., and Linsley, P. S. (2003) Expression profiling reveals off-target gene regulation by RNAi. *Nat. Biotechnol.* **21**, 635–637.
8. Xu, P., Zhang, Y., Kang, L., Roossinck, M. J., and Mysore, K. S. (2006) Computational estimation and experimental verification of off-target silencing during posttranscriptional gene silencing in plants. *Plant Physiol.* **142**, 429–440.
9. Jackson, A. L. and Linsley, P. S. (2004) Noise amidst the silence: off-target effects of siRNAs? *Trends Genet.* **20**, 521–524.
10. Lin, X., Ruan, X., Anderson, M. G., McDowell, J. A., Kroeger, P. E., Fesik, S. W., and Shen, Y. (2005) siRNA-mediated off-target gene silencing triggered by a 7 nt complementation. *Nucl. Acids Res.* **33**, 4527–4535.
11. Qiu, S., Adema, C. M., and Lane, T. (2005) A computational study of off-target effects of RNA interference. *Nucl. Acids Res.* **33**, 1834–1847.
12. Saxena, S., Jónsson, Z., and Dutta, A. (2003) Small RNAs with imperfect match to endogenous mRNA repress translation. *J. Biol. Chem.* **278**, 44312–44319.
13. Dunoyer, P., Himber, C., and Voinnet, O. (2005) DICER-LIKE 4 is required for RNA interference and produces the 21-nucleotide small interfering RNA component of the plant cell-to-cell silencing signal. *Nat. Genet.* **37**, 1356–1360.
14. Agrawal, N., Dasaradhi, P. V. N., Mohmmed, A., Malhotra, P., Bhatnagar, R. K., and Mukherjee, S. K. (2003) RNA interference: biology, mechanism, and applications. *Microbiol. Mol. Biol. Rev.* **67**, 657–685.
15. Meister, G. and Tuschl, T. (2004) Mechanisms of gene silencing by double-stranded RNA. *Nature* **431**, 343–349.
16. Bhaskar, P. B., Venkateshwaran, M., Wu, L., Ane, J.-M., and Jiang, J. (2009) *Agrobacterium*-mediated transient gene expression and silencing: a rapid tool for functional gene assay in potato. *PLoS ONE* **4**, e5812.
17. Higuchi, M., Yoshizumi, T., Kuriyama, T., Hara, H., Akagi, C., Shimada, H., and Matsui, M. (2009) Simple construction of plant RNAi vectors using long oligonucleotides. *J. Plant Res.* **122**, 477–482.
18. Senthil-Kumar, M., Hema, R., Suryachandra, T. R., Ramegowda, H. V., Gopalakrishna, R., Rama, N., Udayakumar, M., and Mysore, K. S. (2010) Functional characterization of three water deficit stress-induced genes in tobacco and *Arabidopsis*: an approach based on gene down regulation. *Plant Physiol. Biochem.* **48**, 35–44.

19. Tang, W., Weidner, D. A., Hu, B. Y., Newton, R. J., and Hu, X.-H. (2006) Efficient delivery of small interfering RNA to plant cells by a nanosecond pulsed laser-induced stress wave for posttranscriptional gene silencing. *Plant Sci.* **171**, 375–381.
20. Warthmann, N., Chen, H., Ossowski, S., Weigel, D., and Herve, P. (2008) Highly specific gene silencing by artificial miRNAs in rice. *PLoS ONE* **3**, e1829.
21. Gheysen, G. and Vanholme, B. (2007) RNAi from plants to nematodes. *Trends Biotechnol.* **25**, 89–92.
22. Wesley, S. V., Christopher, A. H., Neil, A. S., MingBo, W., Dean, T. R., Qing, L., Paul, S. G., Surinder, P. S., David, A., Peter, A. S., Simon, P. R., Andrew, P. G., Allan, G. G., and Peter, M. W. (2001) Construct design for efficient, effective and high-throughput gene silencing in plants. *Plant J.* **27**, 581–590.
23. Filichkin, S. A., DiFazio, S. P., Brunner, A. M., Davis, J. M., Yang, Z. K., Kalluri, U. C., Arias, R. S., Etherington, E., Tuskan, G. A., and Strauss, S. H. (2007) Efficiency of gene silencing in *Arabidopsis*: direct inverted repeats vs. transitive RNAi vectors. *Plant Biotechnol. J.* **5**, 615–626.
24. Stephan, O., Rebecca, S., and Detlef, W. (2008) Gene silencing in plants using artificial microRNAs and other small RNAs. *Plant J.* **53**, 674–690.
25. Carole, L. T., Louise, J., David, C. B., and Andrew, J. M. (2001) Size constraints for targeting post-transcriptional gene silencing and for RNA-directed methylation in *Nicotiana benthamiana* using a *potato virus X* vector. *Plant J.* **25**, 417–425.
26. Bleys, A., Vermeersch, L., Van Houdt, H., and Depicker, A. (2006) The frequency and efficiency of endogene suppression by transitive silencing signals is influenced by the length of sequence homology. *Plant Physiol.* **142**, 788–796.
27. Sijen, T., Fleenor, J., Simmer, F., Thijssen, K. L., Parrish, S., Timmons, L., Plasterk, R. H. A., and Fire, A. (2001) On the role of RNA amplification in dsRNA-triggered gene silencing. *Cell* **107**, 465–476.
28. Kerschen, A., Napoli, C. A., Jorgensen, R. A., and Müller, A. E. (2004) Effectiveness of RNA interference in transgenic plants. *FEBS Lett.* **566**, 223–228.
29. Kohli, A., Twyman, R. M., Abranches, R., Wegel, E., Stoger, E., and Christou, P. (2003) Transgene integration, organization and interaction in plants. *Plant Mol. Biol.* **52**, 247–258.
30. Small, I. (2007) RNAi for revealing and engineering plant gene functions. *Curr. Opin. Biotechnol.* **18**, 148–153.
31. Senthil-Kumar, M., Hema, R., Ajith, A., Li, K., Udayakumar, M., and Kirankumar, S. M. (2007) A systematic study to determine the extent of gene silencing in *Nicotiana benthamiana* and other Solanaceae species when heterologous gene sequences are used for virus-induced gene silencing. *New Phytol.* **176**, 782–791.
32. Jackson, A. L., Burchard, J., Leake, D., Reynolds, A., Schelter, J., Guo, J., Johnson, J. M., Lim, L., Karpilow, J., Nichols, K., Marshall, W., Khvorova, A., and Linsley, P. S. (2006) Position-specific chemical modification of siRNAs reduces "off-target" transcript silencing. *RNA* **12**, 1197–1205.
33. Van Houdt, H., Bleys, A., and Depicker, A. (2003) RNA target sequences promote spreading of RNA silencing. *Plant Physiol.* **131**, 245–253.
34. Dimitri Semizarov, L. F., Aparna, S., Paul, K., Halbert, D. N., and Fesik, S. W. (2003) Specificity of short interfering RNA determined through gene expression signatures. *Proc. Natl. Acad. Sci. USA* **100**, 6347–6352.
35. Khvorova, A., Reynolds, A., and Jayasena, S. D. (2003) Functional siRNAs and miRNAs exhibit strand bias. *Cell* **115**, 209–216.
36. Muckstein, U., Tafer, H., Hackermuller, J., Bernhart, S. H., Stadler, P. F., and Hofacker, I. L. (2006) Thermodynamics of RNA–RNA binding. *Bioinformatics* **22**, 1177–1182.
37. Birmingham, A., Anderson, E. M., Reynolds, A., Ilsley-Tyree, D., Leake, D., Fedorov, Y., Baskerville, S., Maksimova, E., Robinson, K., Karpilow, J., Marshall, W. S., and Khvorova, A. (2006) 3′ UTR seed matches, but not overall identity, are associated with RNAi off-targets. *Nat. Methods* **3**, 199–204.
38. Dhirendra, K., Claes, G., and Daniel, F. K. (2006) Validation of RNAi silencing specificity using synthetic genes: salicylic acid-binding protein 2 is required for innate immunity in plants. *Plant J.* **45**, 863–868.
39. Grimm, D., Streetz, K. L., Jopling, C. L., Storm, T. A., Pandey, K., Davis, C. R., Marion, P., Salazar, F., and Kay, M. A. (2006) Fatality in mice due to oversaturation of cellular microRNA/short hairpin RNA pathways. *Nature* **441**, 537–541.

Chapter 3

Plant Gateway Vectors for RNAi as a Tool for Functional Genomic Studies

Toshiya Muranaka

Abstract

Gateway® system takes advantage of high-throughput creation of various expression vectors from one entry vector. This technology is also applied to RNAi vectors for functional analysis of plant genomics. To date, several plant Gateway vectors have been developed and distributed to plant science community. Here I would like to introduce unique plant Gateway vectors developed for functional analysis of the metabolic pathway in root tissues. The protocol shown here is basically applied to other plant Gateway vectors for RNAi.

Key words: *Agrobacterium rhizogenes*, *Agrobacterium tumefaciens*, functional genomics, Gateway, GFP, hairy root, metabolic regulation.

1. Introduction

Tremendous information on genome sequencing and expressed sequence tag are now available in recent years. Gene knock-out (KO) is useful to understand the function of relevant genes by examining the loss-of-function phenotype. However, because homologous recombination occurs infrequently and random integration predominates in plants, comprehensive KO studies are restricted to model plants such as *Arabidopsis thaliana* and rice in which large tagging libraries are available. Instead, RNA interference (RNAi), a key defense against viruses, as well as a way of regulating endogenous genes in plants can be applied to gene knockdown studies.

Gateway® cloning technology (Invitrogen Co., Ltd.) takes advantage of the site-specific recombination reaction enabling the bacteriophage λ to integrate and excise itself in and out of

H. Kodama, A. Komamine (eds.), *RNAi and Plant Gene Function Analysis*, Methods in Molecular Biology 744, DOI 10.1007/978-1-61779-123-9_3,

bacterial chromosome (1), by which high-throughput creation of various expression vectors from one entry vector is available. To date several plant Gateway vectors have been developed (2–5) and distributed to plant science community.

Culture system of hairy roots, naturally occurring transformed roots, has several desirable features to be an experimental model for functional analyses of the genes involved in root metabolism and/or physiology. We have developed Gateway vector set for efficient target gene RNAi in the transformed roots. The vector pHR-RNAi contains a cluster of *rol* (rooting locus) genes, together with inverted repeats of Gateway conversion cassette separated by an intron sequence, flanked by a dual CaMV 35S promoter and nopaline synthase polyadenylation signal, on the same T-DNA. Model experiment focused on the sterol biosynthetic pathway in *Arabidopsis* validated the utility of pHR-RNAi vectors for gene function analysis in cultured roots (6).

2. Materials

2.1. Culture Medium

1. LB broth; 10 g Bacto tryptone, 5 g Bacto yeast extract, 5 g NaCl, 2 mL of 1 M NaOH in 1 L of distilled water (*see* **Note 1**).
2. SOC broth: 20 g Bacto tryptone, 5 g Bacto yeast extract, 0.6 g NaCl, 0.5 g KCl in 1 L distilled water. Autoclave, cool, then add 10 mL of 1 M $MgCl_2$, 10 mL of 1 M $MgSO_4$, and 20 mL of 20 mM glucose.
3. LB agar plates containing 100 μg/mL spectinomycin.
4. LB plates containing 100 μg/mL kanamycin.
5. YEB medium: 5 g Bacto peptone, 5 g beef extract, 1 g Bacto yeast extract, 5 g sucrose, 2 mM $MgSO_4$ in 1 L of distilled water, adjust to pH 7.2 by 1 N NaOH.

2.2. Molecular Cloning Kits and Agents

1. TaKaRa Ex Taq® (Takara, Japan) (*see* **Note 2**).
2. SMT-F primer; 5′-ACC*CTCGAG*TACGAGTGGGGATG GGGACAATC-3′, SMT-R primer; 5′-GCT*GGATCC* CAGAGAATCATATGCATCGGAG-3′ (*see* **Note 3**).
3. Full length of *Arabidopsis* SMT2 and SMT3 cDNAs (*see* **Note 4**).
4. Restriction enzymes: *Xho*I, *Bam*HI (Takara, Japan).
5. pENTR®1A vector (Invitrogen).
6. T4 DNA ligase and 10× ligation buffer.
7. One Shot® chemically competent *Escherichia coli* (Invitrogen).

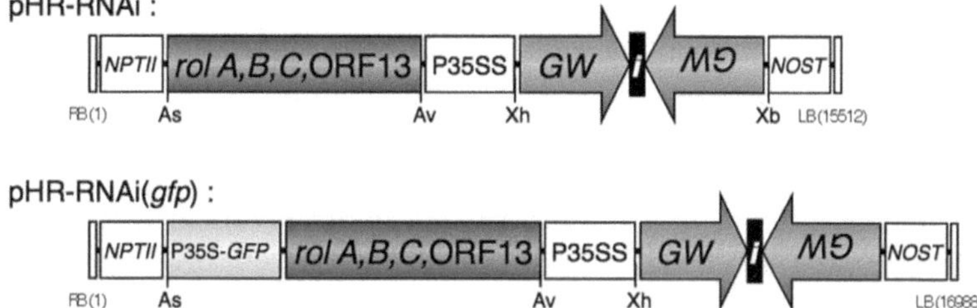

Fig. 3.1. Vector structure. Only the region between the *right* and *left* borders is shown (not to scale). *NPTII*, neomycin phosphotransferase II gene; P35S-*GFP*, *sGFP*(S65T) coding sequence flanked by the CaMV 35S promoter and nopaline synthase polyadenylation signal; *rol A, B, C,* ORF13, 7.6-kb fragment derived from pRi1724 in *Agrobacterium rhizogenes* strain MAFF03-01724 (15); P35SS, CaMV 35S promoter with duplicated enhancer region; *GW*, GATEWAY^{TM} conversion cassette (*attR1, Cm^R, ccdB, attR2* orientation); *i*, a 250-bp intron sequence derived from *Arabidopsis WRKY33* gene (*At2g38470*); *NOST*, nopaline synthase polyadenylation signal; RB and LB, T-DNA *right* and *left borders*, respectively. *Numbers* in parenthesis indicate the positions relative to the *right* T-DNA border. Several of the unique sites on T-DNA are shown, *Asc*I (As), *Avr*II (Av), *Xho*I (Xh), *Xba*I (Xb). (Reproduced from ref. 6 with permission from Japanese Society for Plant Cell and Molecular Biology.)

8. Spectinomycin stock solution (100 mg/mL, 1000×).
9. TE buffer: 10 mM Tris–HCl, pH 8.0, 1 mM Na_2EDTA.
10. LR Clonase®II enzyme mix (Invitrogen).
11. Plant Gateway vectors for RNAi: pHR-RNAi, pHR-RNAi(*gfp*) (**Fig. 3.1**) (6) (*see* **Note 5**).
12. Proteinase K solution: 20 mg/mL proteinase K (Invitrogen).
13. One Shot® OmniMAX® 2 T1 Phage-Resistant Cell (Invitrogen).
14. Kanamycin stock solution (100 mg/mL, 1000×).

2.3. Plant Transformation

1. *Agrobacterium tumefaciens* strain GV3101(pMP90) (7).
2. 10% Glycerol.
3. Gene Pulser Xcell (Bio-rad).
4. *Arabidopsis thaliana* (Ecotype Colombia or WS).
5. 5% Sucrose solution.
6. Silwet L-77.

3. Methods

Here shows model experiments to knock down the gene for the sterol biosynthetic pathway in *Arabidopsis*. At first, cDNA fragments (here SMT2 and SMT3 cDNA) were cloned into

Gateway entry vector and then plant binary vector was created by LR recombination. After that, *Arabidopsis* was transformed by simple floral dipping method. This protocol can be basically applied to other plant Gateway vectors for RNAi.

3.1. Creation of a Gateway Entry Clone

1. Produce PCR products using Taq DNA polymerase (TaKaRa Ex Taq®) and primer set (here SMT-F and MST-R) using full-length SMT2 or SMT3 cDNA as a template, respectively.
2. Digest the PCR products with *Xho*I and *Bam*HI.
3. Digest pENTR®1A vector with *Xho*I and *Bam*HI.
4. Ligate the PCR product with *Xho*I/*Bam*HI-digested pENTR®1A vector.
5. Add 2 L of the above ligation sample into a vial of One Shot® chemically competent *E. coli* and mix gently.
6. Incubate on ice for 5 min.
7. Heat shock the cells for 30 s at 42°C without shaking. Immediately transfer the tube to ice.
8. Add 250 L of room temperature SOC medium.
9. Incubate at 37°C for 1 h with shaking.
10. Spread 10–50 μL of bacterial culture on a prewarmed LB agar plate containing 100 μg/mL spectinomycin and incubate overnight at 37°C.

3.2. LR Recombination Between pHR-RNAi(gfp) and an Entry Clone

1. Add 1–7 μL of entry clone (50–150 ng) constructed in **Section 3.1** and 1 μL of pHR-RNAi(*gfp*) (150 ng/μL) into a 1.5-mL tube. Bring to 8 μL with TE buffer at room temperature and mix.
2. Thaw on ice the LR Clonease®II enzyme for about 2 min. Vortex the LR Clonase™II enzyme briefly twice (2 s each time).
3. Add 2 μL of LR Clonase®II enzyme mix to the reaction (from **Section 3.2**, step 1) and mix well by vortexing briefly twice. Microcentrifuge briefly.
4. Incubate reactions at 25°C for 1 h.
5. Add 1 μL of the proteinase K solution to each sample to terminate the reaction. Vortex briefly. Incubate sample at 37°C for 10 min.
6. Add 1 μL of each LR reaction into 50 μL of One Shot® OmniMAX® 2 T1 Phage-Resistant Cells. Incubate on ice for 30 min. Heat shock cells by incubating at 42°C for 30 s.
7. Add 250 μL of SOC Medium and incubate at 37°C for 1 h with shaking. Plate 20 and 100 μL of transforma-

tion onto selective plates. Transform 1 μL of pUC19 DNA (10 ng/μL) into 50 μL of One Shot® OmniMAX™ 2 T1 Phage-Resistant Cells as described above. Plate 20 and 100 μL on LB plates containing 100 μg/mL kanamycin, or the appropriate selection marker for your donor vector (*see* **Note 6**).

3.3. Introduction of Binary Vectors into Agrobacterium tumefaciens by Electroporation

1. Grow *A. tumefaciens* strain GV3101(pMP90) in 5 mL of liquid LB medium overnight at 30°C (8).
2. Transfer 0.5 mL of cultured *A. tumefaciens* obtained from step 1 to 50 mL of liquid YEB medium for 6–9 h at 30°C. Allow the culture to attain a density of 1–2 × 10^8 cells/mL.
3. Transfer the culture broth to the centrifuge tube and keep on ice. Sediment the bacteria by spinning at 5,000×*g* for 5 min at 4°C. Decant the supernatant and suspend the pellet in 20 mL of ice-cold 10% glycerol. Repeat this step at least three times to remove the culture broth completely.
4. After centrifugation, suspend the pellet in 125 μL of 10% glycerol. Divide each 40 L of the suspension into a 1.5-mL centrifugation tube. Keep the samples in the freezer at –80°C until use. They can be kept for 1–2 years.
5. Thaw the suspension cells on ice and add 1 μL of 50 ng/L binary vector to the tube. Mix gently with an ice-cold sterile glass pipette and transfer to an ice-cold 0.2-cm cuvette.
6. Set the Gene Pulser apparatus at 20 mF. Set the Pulse Controller to 200 Ω and the Gene Pulser apparatus to 2.5 kV.
7. Pulse once at the above setting and quickly transfer the suspension to a culture tube containing 1 mL of YEB. Incubate for 1 h at 30°C.
8. Dilute the culture broth with YEB. Streak 100 μL of 1/100-diluted cells on LB plate (1.5% agar) containing suitable antibiotics (pHR-RNAi(gfp) vector; 200 g/mL kanamycin). Incubate at 30°C for 2–3 days. Usually more than 100 colonies result.
9. Pick up single colony and streak on the LB plate with suitable antibiotics (at step 8) for single-cell purification.
10. To confirm the introduction of the binary vector, prepare plasmid DNA from the transformed *A. tumefaciens* cells by alkaline lysis.

3.4. Transformation of Arabidopsis

This section is modified from ref. (9).

1. Grow *Arabidopsis* plants in pots in soil covered with window screen under 16-h light and 8-h dark condition until they are flowering.

2. Clip off the first bolts to enhance the proliferation of many secondary bolts.
3. Culture *A. tumefaciens* (prepared in **Section 3.3**) in 5 mL of liquid LB medium with 200 g/mL kanamycin until mid-log cells or a recently stationary culture.
4. Spin down *Agrobacterium*, resuspend to $OD_{600} = 0.8$ in 5% sucrose solution and add 500 L of Silwet L-77 per liter of sucrose solution (*see* **Note 7**).
5. Dip above-ground parts of plant in *Agrobacterium* solution for 2–3 s with gentle agitation and place dipped plants laid on their side with covering rapping film for 24 h to maintain high humidity (*see* **Note 8**).
6. Water and grow plants normally, tying up loose bolts with twist ties. Stop watering as seeds become mature.
7. Harvest dry seeds (*see* **Note 9**). Select for transformant on agar medium containing 50 mg/L of kanamycin (*see* **Note 10**). Separate a root cluster from a seedling and transfer to liquid medium, or transplant the seedling to soil and allow to continue growing.

4. Notes

1. Culture medium is autoclaved at 121°C for 15 min.
2. TaKaRa Ex Taq® or equivalent DNA polymerases commercially available can be used.
3. In this protocol, for the purpose of RNAi of sterol metabolite regulation in root tissue, 800-bp fragment within the coding sequence of each of *SMT2* and *SMT3* genes (**Fig. 3.2**) was used. Underlined sequences in each of the primers corresponded to the restriction enzyme sites (*Xho*I and *Bam*HI, respectively) for cloning purposes. The amplified region included 253–1,052 bp of the SMT2 and SMT3 cDNAs.
4. Various *Arabidopsis* full-length cDNAs are obtained from RIKEN Bioresource Center (http://www.brc.riken.go.jp/lab/epd/Eng/catalog/RAFL.shtml) (10–12).
5. Both vectors are available from Toshiya Muranaka, Osaka University (muranaka@bio.eng.osaka-u.ac.jp). For other plant Gateway vectors for RNAi, should be asked to the corresponding authors of each publication (2–5).
6. Any competent cells with a transformation efficiency of $>1.0 \times 10^8$ transformants/μg may be used.

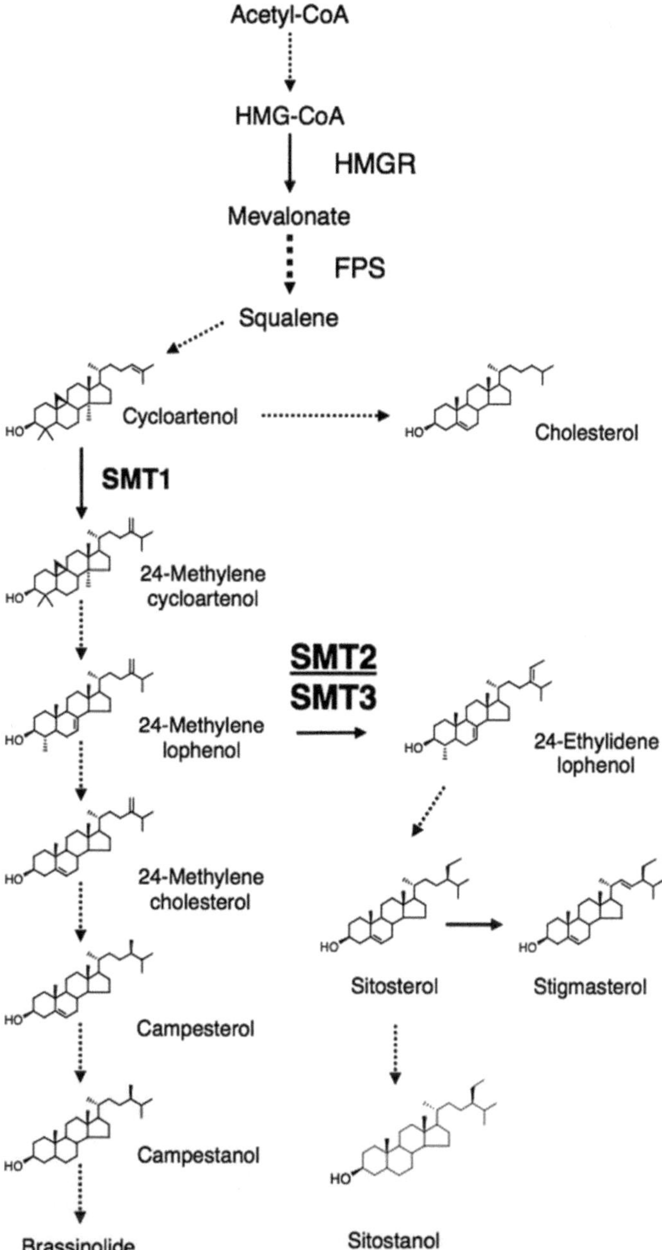

Fig. 3.2. Simplified biosynthetic pathway for plant sterols. *Solid arrows* indicate a single enzymatic step, and *dashed arrows* indicate more than one enzymatic step. HMGR, 3-hydroxy-3-methylglutaryl CoA reductase; FPS, farnesyl diphosphate synthase; SMT, sterol-C24-methyltransferase. (Reproduced from ref. 6 with permission from Japanese Society for Plant Cell and Molecular Biology.)

7. You will need 100–200 mL for each two or three small pots to be dipped, or 400–500 mL for each two or three 9-cm pots.
8. Do not expose to excessive sunlight to avoid getting hot.
9. Transformant are usually all independent but are guaranteed to be independent if they come off of separate plants.
10. To establish root cultures, cluster of roots was separated from individual seedling (2-week-old) and transferred to hormone-free liquid Gamborg B5 medium (13) or MS medium (14). We confirmed that established root cultures could be maintained at least for 2 years without loss of GFP fluorescence by subculturing every 2–3 weeks. Northern analyses of root culture lines transformed with *SMT*–RNAi constructs confirmed that all of the lines analyzed showed decreased transcript levels of corresponding gene compared with those in the empty vector lines (**Fig. 3.3**). Sterol profile in *SMT*–RNAi-transformed hairy roots was altered as expected (detailed in ref. 6).

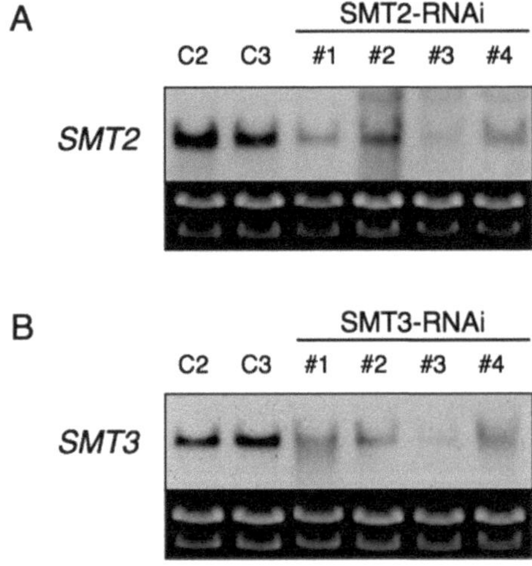

Fig. 3.3. Northern analysis of *SMT*–RNAi lines. **a** Expression of *SMT2* in empty vector lines (C2 and C3) and four independent *SMT2* RNAi lines (SMT2-RNAi). **b** Expression of *SMT3* in empty vector lines (C2 and C3) and four independent *SMT3* RNAi *lines* (SMT3-RNAi). Ethidium bromide staining of rRNAs is shown to demonstrate equal loading of RNA in each lane. (Reproduced from ref. 6 with permission from Japanese Society for Plant Cell and Molecular Biology.)

Acknowledgment

This work was supported in part by grants in aid for scientific research from The Ministry of Education, Culture, Sports, Science and Technology of Japan.

References

1. Katzen, F. (2007) Gateway recombinational cloning: a biological operating system. *Expert Opin. Drug. Discov.* **2**, 571–589.
2. Karimi, M., Inze, D., and Depicker, A. (2002) GATEWAY vectors for *Agrobacterium*-mediated plant transformation. *Trends Plant Sci.* **7**, 193–195.
3. Helliwell, C. and Waterhouse, P. (2003) Constructs and methods for high-throughput gene silencing in plants. *Methods* **30**, 289–295.
4. Curtis, M. D. and Grossniklaus, U. (2003) A gateway cloning vector set for high-throughput functional analysis of genes in planta. *Plant Physiol.* **133**, 462–469.
5. Lo, C., Wang, N., and Lam, E. (2005) Inducible double-stranded RNA expression activates reversible transcript turnover and stable translational suppression of a target gene in transgenic tobacco. *FEBS Lett.* **579**, 1498–1502.
6. Seki, H., Ohyama, K., Nishizawa, T., Yoshida, S., and Muranaka, T. (2008) The "all-in-one" rol-type binary vectors as a tool for functional genomic studies using hairy roots. *Plant Biotechnol.* **25**, 347–355.
7. An, G., Ebert, P., Mitra, A., and Ha, S. (1988) Binary vectors. In: *Plant Molecular Biology Manual* (eds. Gelvin, S. B. and Schilperoort, R. A.), Kluwer Academic Publishers, Dordrecht, pp. A3: 1–19.
8. Muranaka, T., Kitamura, Y., and Ikenaga, T. (1999) Genetic transformation of *Duboisia* species. In: *Biotechnology in Agriculture and Forestry. Vol. 45 Transgenic Medicinal Plants* (ed. Bajaj, Y. P. S.), Springer, Berlin, Heidelberg, pp. 117–132.
9. Clough, S. J. and Bent, A. F. (1998) Floral dip: a simplified method for *Agrobacterium*-mediated transformation of *Arabidopsis thaliana*. *Plant J.* **16**, 735–743.
10. Seki, M., Carninci, P., Nishiyama, Y., Hayashizaki, Y., and Shinozaki, K. (1998) High-efficiency cloning of *Arabidopsis* full-length cDNA by biotinylated CAP trapper. *Plant J.* **15**, 707–720.
11. Seki, M., Narusaka, M., Kamiya, A., Ishida, J., Satou, M., Sakurai, T., Nakajima, M., Enju, A., Akiyama, K., Oono, Y., Muramatsu, M., Hayashizaki, Y., Kawai, J., Carninci, P., Itoh, M., Ishii, Y., Arakawa, T., Shibata, K., Shinagawa, A., and Shinozaki, K. (2002) Functional annotation of a full-length *Arabidopsis* cDNA collection. *Science* **296**, 141–145.
12. Sakurai, T., Satou, M., Akiyama, K., Iida, K., Seki, M., Kuromori, T., Ito, T., Konagaya, A., Toyoda, T., and Shinozaki, K. (2005) RARGE: a large-scale database of RIKEN *Arabidopsis* resources ranging from transcriptome to phenome. *Nucleic Acids Res.* **33**(Database issue), D647–D650.
13. Gamborg, O. L., Miller, R. A., and Ojima, K. (1986) Nutrient requirements of suspension cultures of soybean root cells. *Exp. Cell. Res.* **50**, 151–158.
14. Murashige, T. and Skoog, F. (1962) A revised medium for rapid growth and bioassays with tobacco culture. *Physiol. Plant.* **15**, 473–497.
15. Tanaka, N., Ikeda, T., and Oka, A. (1994) Nucleotide sequence of the *rol* region of the mikimopine-type root inducing plasmid pRi1724. *Biosci. Biotechnol. Biochem.* **58**, 548–551.

Chapter 4

Heat-Inducible RNAi for Gene Functional Analysis in Plants

Frédéric Masclaux and Jean-Philippe Galaud

Abstract

Controlling gene expression during plant development is an efficient method to explore gene function and RNA interference (RNAi) is now considered as a powerful technology for gene functional analysis. However, constitutive gene silencing cannot be used with genes involved in fundamental processes such as embryo viability or plant growth and alternative silencing strategies avoiding these limitations should be preferred. Tissue-specific and inducible promoters, able to control gene expression at spatial and/or temporal level, can be used to circumvent viability problems. In this chapter, after a rapid overview of the inducible promoters currently used for transgenic approaches in plants, we describe a method we have developed to study gene function by heat-inducible RNAi. This system is easy to use and complementary to those based on chemical gene inducer treatments and might be useful for both research and biotechnological applications.

Key words: Conditional gene expression, constitutive expression, gene function analysis, heat-shock, *HSP18.2* promoter, inducible promoter, reverse genetics, RNA interference.

1. Introduction

1.1. RNAi, an Efficient Method for Gene Functional Analysis

Arabidopsis thaliana genome sequencing program provided the plant research community with the sequence of about 28,000 genes (1) and the recent development of ultrahigh-throughput sequencing methods (2), able to yield millions of sequences in a single run, will continue to produce a large amount of sequence information on plant genomes in the next future. However, for a large part of genes that were identified, no function has been yet associated and assigning a function for each predicted gene will be a major challenge in the next years. The study of plants in which a gene was inactivated makes it possible to identify the function of this gene and those mutant plants can be found in

H. Kodama, A. Komamine (eds.), *RNAi and Plant Gene Function Analysis*, Methods in Molecular Biology 744,
DOI 10.1007/978-1-61779-123-9_4,

collections of insertion-tagged mutant lines that were created for functional genomic analyses; however, mutant lines are not available for every gene and development of alternative technologies is often necessary. Alternatively, gene expression can be modulated by transgenic approaches using sense, antisense, and RNA interference (RNAi) constructs (3, 4) and RNAi was shown to be a very efficient strategy to downregulate gene expression (3, 5). RNAi is a natural process of gene regulation common to most eukaryotic organisms, which leads to sequence-specific degradation of mRNA in the cytoplasm. This process is widely used for gene downregulation in various organisms such as mammals (6), nematodes (7), and plants (3, 8). The mechanism involved is reliably initiated by double-stranded RNA (dsRNA), which is cleaved by an RNase III-like enzyme complex (Dicer) to produce small interfering RNAs (siRNAs) of 21–25 nucleotides (9). siRNAs then serve as guide sequences for target mRNA degradation. Interestingly, it was also shown that gene suppression can also be achieved by expression of double-stranded RNA designed from promoter sequences. Expression of those RNAs can induce transcriptional silencing that could be stable and inherited (10, 11). More recently, use of artificial microRNAs (amiRNAs) that are 21-mer small RNAs was reported to be an efficient technology to interfere specifically with a single or multiple genes of interest in plants. These amiRNAs can be genetically designed and transferred by transgenic approaches (12). They act as negative regulators on gene expression by cleavage-induced mRNA degradation. A Web tool (http://www.openbiosystems.com/RNAi/ArabidopsisthalianaamiRNA) has been recently developed for the entire *Arabidopsis* genome to help *Arabidopsis* community to identify optimal amiRNA sequences against their favourite gene (12).

1.2. Inducible Promoters

Constitutive promoters such as the cauliflower mosaic virus promoter (35S promoter) are commonly used for plant transformation, but overall activity of this promoter may be responsible for pleiotropic effects on plant development. The main problem encountered to modulate gene expression in transgenic plants is observed when the target gene encodes a key element in basic cell functions, or at particular developmental stages. In those cases, constitutive silencing or constitutive overexpression of the gene may induce plant lethality. Hence, use of promoters that can be induced by physical, chemical, or biological conditions or promoters that are tissue-, organ-, or development specific can be very useful. For instance, a guard cell promoter was recently identified and can be a very elegant and efficient tool to target gene expression or gene silencing in stomata (13). The objective is then to generate RNAi constructs that can be controlled and activated only where and when induction of gene silencing is required. Such constructs, when expressed in plants, will greatly facilitate the analysis of genes whose mutations are lethal or are responsible for a very severe phenotype. Promoters responding to chemical (14) or physical treatments can be used for fundamental research as well as biotechnological applications. With those promoters, transformed plants can be regenerated, while the promoter is inactive and further analysis can be performed when required after activation of the transgene expression by applying the inductive treatment.

Among the inducible expression systems used in plants, several promoters are controlled by various external inducers such as tetracycline (15), copper (16), ethanol (17), and steroid (18). Chemical inducers are easy to apply on plants and are effective at a low concentration and in a dose-dependent manner. Their effects are reversible when the chemical treatment is stopped. However, it was reported that the chemical application can produce growth disturbances as shown for the glucocorticoid-inducible system in the legume *Lotus japonicus* (19) or for the ethanol-inducible gene expression that can impact growth of potatoes plants (20). Furthermore, the absorption and the mobility of the chemical can be a limiting factor for an efficient induction. All together, this indicates that we should be careful with any of the inducible system using chemical treatment.

One of the simplest methods to obtain inducible gene expression is to use a promoter of heat-shock endogenous genes. Heat shock can be applied to many plants at laboratory scale and at different developmental stages. Previous studies on the promoter of the *HSP18.2* gene (*At5g59720*) from *Arabidopsis*, encoding a heat-shock protein, indicated that it functions as a strong inducible system in plants (21–23). *HSP18.2* promoter was not activated at normal growth temperature (25°C) and a 1,000-fold activation is observed following a short heat-shock period at 37°C (24). It can be used not only for *Arabidopsis* transformation but also with good efficiency in others species such as tobacco (24, 25). It can also be employed for other application than RNAi and an interesting work recently reported has shown that *Arabidopsis HSP18.2* promoter can control the excision of a selectable marker from transgenic tobacco after a heat-shock treatment (25).

We described previously the silencing of the phytoene desaturase (*PDS*) gene using a heat-inducible, gene-silencing system in *Arabidopsis* (26). mRNA degradation of *PDS* prevents the synthesis of carotenoids, resulting in a photobleaching phenotype due to chlorophyll photo-oxidation (27, 28). Application of heat treatment to plants containing the inducible gene-silencing system leads to photobleaching in the young organs. Our system is based on an RNAi cassette that contains an excisable intron in order to

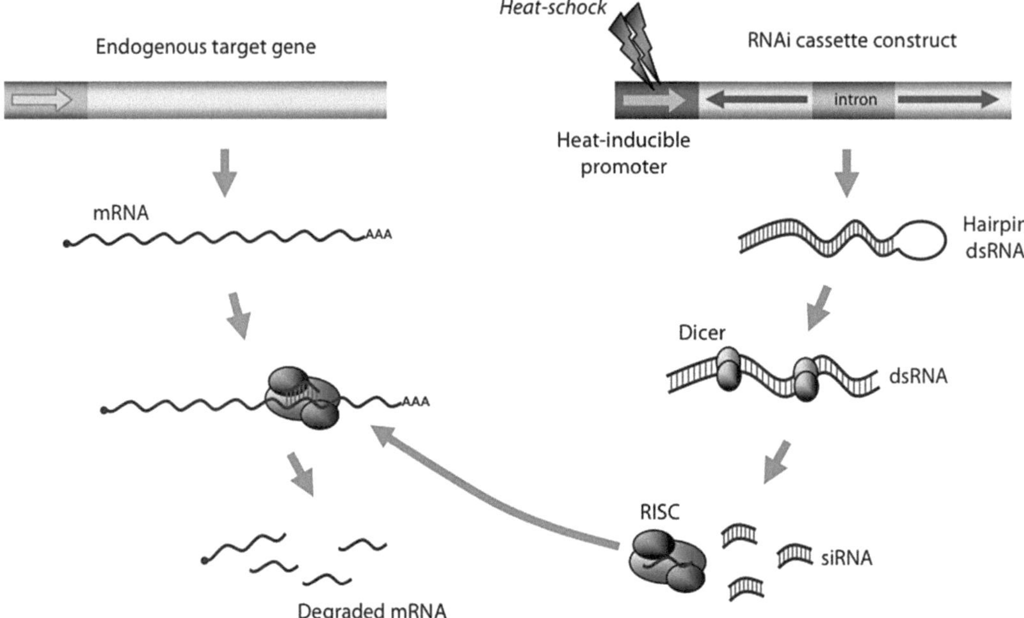

Fig. 4.1. Activation of the *HSP18.2* promoter by heat treatment leads to the production of an auto-complementary mRNA, which will form a hairpin double-stranded RNA (dsRNA). The dsRNA initiates RNAi by activating the ribonuclease protein Dicer, which binds and cleaves double-stranded RNAs (dsRNAs) to produce double-stranded fragments of 21–25 base pairs. These short double-stranded fragments are called small interfering RNAs (siRNAs). These siRNAs are then separated into single strands and integrated into an active RISC complex. After integration into the RISC, siRNAs base-pair to their target mRNA and induce cleavage of the mRNA, thereby preventing it from being used as a translation template. Consequently, the mRNA of the gene of interest is degraded by the RNAi cassette products upon heat treatment.

facilitate hairpin formation and to enhance the silencing efficiency as previously shown (29). This RNAi cassette is controlled by the *HSP18.2* heat-shock promoter. The general mechanism is summarized in **Fig. 4.1**. This chapter aims to describe the method to use our system for heat-inducible gene silencing in *Arabidopsis*. We describe our own strategy to generate and use heat-inducible RNAi in *Arabidopsis*, but the vectors and the cloning strategy can be divergent from our system.

First, the strategy to select the target region in the gene of interest and the cloning procedure are described. The RNAi cassette is composed of three parts (1): the left arm which is the antisense of the target region in the gene of interest (2); the linker which is an excisable intron (3); and the right arm which is the sense of the target region in the gene of interest and which is strictly complementary to the left arm. This cassette is transferred to a binary vector containing the heat-inducible promoter. Second, the methods to obtain and select transgenic plants are explained. Finally, we detail the different approaches to induce gene silencing by heat treatment.

2. Materials

2.1. Selection of a Target Region in the Gene of Interest and Cloning

1. Polymerase chain reaction (PCR) primers.
2. High-fidelity Taq polymerase (5 U/μL) and its dedicated buffer.
3. 10 mM dNTPs each. Store at –20°C.
4. Agarose.
5. 0.2 mg/mL Ethidium bromide (EtBr) solution. Store at 4°C and protect from light. This product is highly toxic and care should be taken.
6. TAE buffer (50× stock): 2 M Tris, 1 M glacial acetic acid, 50 mM Na_2EDTA, pH 8.0.
7. TE buffer: 10 mM Tris–HCl, pH 8.0, 1 mM Na_2EDTA.
8. DNA loading buffer: 15 g Ficoll 400, 0.5 g sodium dodecyl sulphate (SDS), 0.25 g bromophenol blue. Bring to 100 mL with TE buffer. Store at 4°C.
9. Gel extraction kit (e.g. QIAEX II Gel Extraction Kit), if required.
10. TA-vector cloning kit (e.g. Promega pGEM-T Easy Vector System).
11. *Escherichia coli* cells (e.g. DH5α). Store at –80°C.
12. SOC broth: 20.0 g Bacto tryptone, 5.0 g Bacto yeast extract, 0.6 g NaCl, 0.5 g KCl. Bring to 1 L with deionized water. Autoclave, cool, then add 10 mL of 1 M $MgCl_2$, 10 mL of 1 M $MgSO_4$ and 20 mL of 20 mM glucose (all three solutions sterilized through a 0.45-μm disposable filter).
13. Luria–Bertani (LB) broth: 10.0 g bacteriological tryptone (Difco), 5.0 g yeast extract, 10.0 g NaCl. Bring to 1 L with deionized water and adjust pH to 7 using 1 M NaOH.
14. Ampicillin stock solution (50 mg/mL, 1000×): 0.25 g ampicillin; bring to 5 mL with deionized water, sterile-filter. Store at –20°C.
15. LB–ampicillin plates: 15.0 g Bacto agar. Bring to 1 L with LB broth, autoclave, and cool in a water bath to 50°C. Add 1 mL of ampicillin stock solution and pour the plates. Store for a few weeks at 4°C.
16. Glass beads (4 mm diameter).
17. Plasmid miniprep kit (e.g. Qiagen QIAprep Spin Miniprep Kit).
18. Sequencing primers corresponding to the TA vector.

2.2. Construction of the RNAi Vectors for Plant Transformation

1. *See* **Section 2.1** for agarose, ethidium bromide (EtBr) solution, TAE buffer, DNA loading buffer, Gel extraction kit, *E. coli*-competent cells, SOC broth, LB agar plates with ampicillin, LB broth, ampicillin, glass beads, plasmid miniprep kit.
2. Restriction nucleases: *Hin*dIII, *Sal*I, *Xho*I, *Eco*RI, *Spe*I, *Bam*HI, and *Xba*I.
3. Vectors pBSint11, pGind01, pSR01. PBSint11 is based on the pBluescript II vector background. pGind01 and pSR01 are pGreen vector derivatives (30). Vectors are available on request and their maps are presented in **Fig. 4.2**.
4. T4 DNA ligase and 10× ligation buffer.
5. DNA cleanup kit (e.g. QIAEX II Gel Extraction Kit that is also adapted for this purpose).

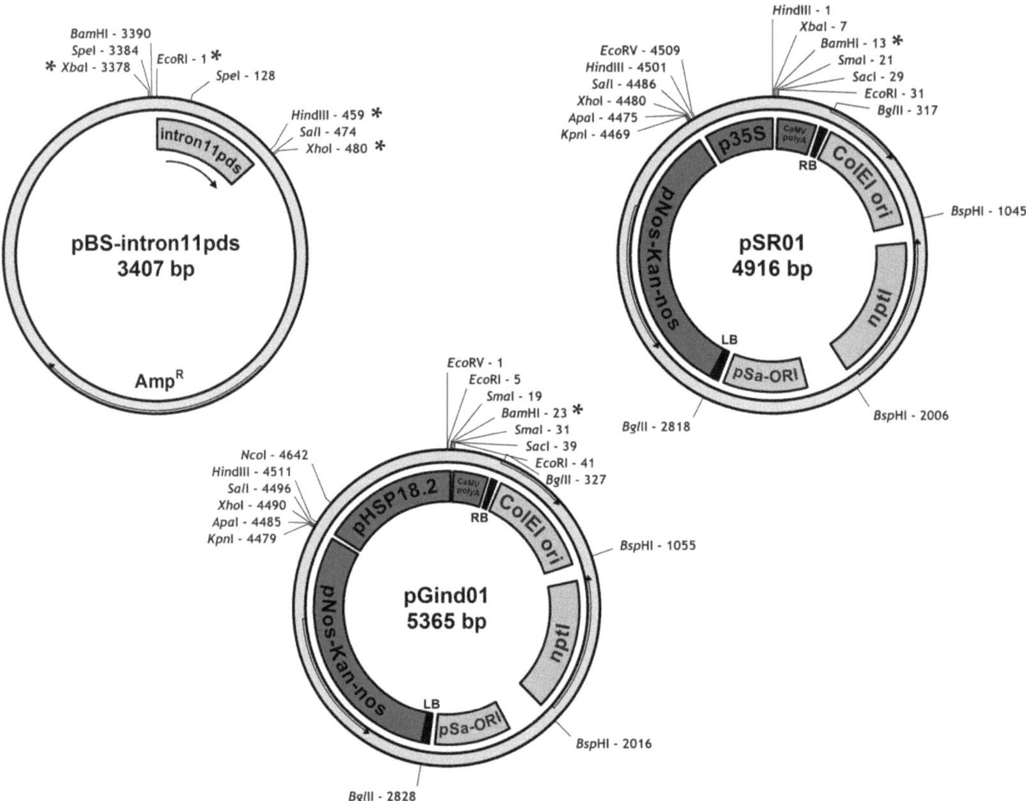

Fig. 4.2. Vector maps. pBSint11 is a pBluescript II-derivative vector containing the intron no. 11 from the phytoene desaturase gene. This vector is used to construct the RNAi cassette. pSR01 is a pGreen-derivative vector containing a 35S promoter and a CAMV polyA to allow constitutive expression of the transgene. pGind01 is a pGreen-derivative vector similar to the pSR01 but containing the heat-inducible *HSP18.2* promoter instead of the 35S promoter. Restriction sites are presented for the most frequent enzymes and the *stars* mark the restriction sites used in the cloning procedure. Vector maps were generated using the program pDraw, available from www.acaclone.com. Complete sequences available on request.

6. Shrimp alkaline phosphatase (SAP) and its corresponding buffer (e.g. Fermentas).
7. Kanamycin stock solution (50 mg/mL, 1000×): 0.25 g kanamycin; bring to 5 mL with deionized water, sterile-filter. Store at –20°C. Care should be taken during manipulation because of its toxicity.
8. LB agar plates containing 50 μg/mL kanamycin. Store for a few weeks at 4°C.

2.3. Transformation of Agrobacterium with the RNAi Vector

1. *Agrobacterium tumefaciens* strain GV3101::pMP90, also called C58C1 RifR::pMP90 (31).
2. YEP medium: 10 g yeast extract, 10 g Bacto peptone, 5 g NaCl. Adjust pH to 7.0 and bring final volume to 1 L with deionized water. Autoclave.
3. Gentamicin stock solution (20 mg/mL, 1000×): 0.1 g gentamicin; bring to 5 mL with deionized water, sterile-filter. Store at –20°C. Care should be taken during manipulation because of its toxicity.
4. Rifampicin stock solution (50 mg/mL, 1000×): 0.25 g rifampicin; bring to 5 mL with DMSO. Store at 4°C. Care should be taken during manipulation because of its toxicity.
5. 20 mM $CaCl_2$.
6. Glycerol.
7. Vector pSOUP; the pGreen system requires this additional helper plasmid (30).
8. YEP agar plates containing 20 μg/mL gentamicin, 50 μg/mL rifampicin, and 50 μg/mL kanamycin: 15.0 g Bacto agar. Bring to 1 L with YEP medium, autoclave, and cool in a water bath to 50°C. Add 1 mL of each antibiotic stock solution and pour the plates. Store for a few weeks at 4°C.
9. Glass beads (4 mm diameter).
10. Plasmid miniprep kit (e.g. Qiagen QIAprep Spin Miniprep Kit).
11. Restriction nucleases: *Bam*HI.

2.4. Transformation of Arabidopsis Plants

1. *Arabidopsis thaliana* seeds (WT or mutant depending on interest).
2. Plastic pots and soil.
3. Standard fertilizer.
4. Nylon mesh (if required).
5. YEP medium (*see* **Section 2.3**, item 2).

6. Antibiotics stock solution: gentamicin, rifampicin, and kanamycin (*see* **Section 2.3**, item 3; **Section 2.3**, item 4; **Section 2.2**, item 7, respectively).
7. Infiltration medium: 0.5× Murashige and Skoog salts; adjust pH to 5.7 with 1 M KOH; add 5% sucrose (w/v), 0.04% Silwet L-77 (also known as "Vac-in-Stuff", Lehle Seeds). No sterilization required, use immediately.
8. Saran wrap or a box with cover.

2.5. Selection of the RNAi Transgenic Plants

1. Sterilization solution: 1 volume of chlorine solution at 0.4 chlorometric degree (dilute bleach to get 2.4% active chlorine) and 5 volumes of 95% ethanol. Keep in dark not longer than 1 month.
2. 95% Ethanol
3. Sterile filters
4. Half-strength MS medium: 0.5× Murashige and Skoog salts; adjust pH to 5.7 with 1 M KOH; autoclave directly or proceed with phytoagar to make plates.
5. Agar plates containing half-strength MS medium supplemented with 50 μg/mL kanamycin. Add 0.8 g/L of phytoagar to half-strength MS medium. Autoclave and cool in a water bath to 50°C. Add 1 mL of kanamycin stock solution and pour the plates. Store a few weeks at 4°C.
6. Microporous tape.

2.6. Heat Induction of the RNAi Construct and Plant Characterization

1. Agar plates containing half-strength MS medium supplemented with 50 μg/mL kanamycin (*see* **Section 2.5**, item 5).
2. Microporous tape.
3. Plastic pots and soil.
4. Thermometer.
5. Ultrasonic fog generator, also named "mist maker," composed of one vibrating ceramic disc. Can be found easily in garden shops at a low price.
6. Heat-shock chamber composed of a large box with a cover, an ultrasonic fog generator, and a beaker filled with water. This chamber is placed in a standard incubator at 37°C (like a bacterial incubator or a phytotron) prior to the experiment (warming up). The ultrasonic fog generator is dipped in the beaker. Once the plants are added, the chamber is tightly closed to avoid humidity escape and the fog generator is started. The fog generator is placed under the control of a timer plug with a cycle of 5 min on/10 min off. This system produces a highly wet atmosphere at 37°C.
7. RNA extraction kit (e.g. Qiagen RNeasy Mini kit).

3. Methods

3.1. Selection of a Target Region in the Gene of Interest and Cloning

1. Selection of a target region is crucial for effective gene silencing of the gene of interest. The target region should correspond to exons but target region with two or more exons interrupted by introns will also work well. We generally aim for ~300–600-bp products, although RNAi with products ranging from 50 to 3,000 bp has been shown to work.
2. The target region should not contain complete 19-mer homology to other genes or the produced dsRNA could be non-specific (*see* **Note 1**). When a gene presents a strong identity with other genes (in case of a multigenic family, for instance), the 3′ UTR can be used as a specific target region.
3. Define standard primers (20-mer) at each extremity of the target region. The target region will be amplified with two sets of primers to generate two identical fragments (A1 and A2) except for the extremities harbouring different restrictions sites.
4. For the first set of primers: The sequence TA*AAGCTT* is added at the 5′-end of the forward primer to insert the *Hin*dIII restriction site. For the reverse primer, the sequence TA*GTCGACGGATCC* is added at the 5′-end to insert the *Sal*I and *Bam*HI restriction sites. The selected restriction sites have to be absent from the target region. The fragment cloned using this set is named "A1."
5. For the second set of primers: The sequence TA*GAATTC* is added at the 5′-end of the forward primer to insert the *Eco*RI restriction site. For the reverse primer, the sequence TA*ACTAGTGGATCC* is added at the 5′-end to insert the *Spe*I and *Bam*HI restriction sites. The selected restriction sites have to be absent from the target region. The fragment cloned using this set is named "A2."
6. Perform two standard 50 μL PCR reactions using your selected primer sets to amplify the fragment of interest A1 and A2. Use a high-fidelity DNA polymerase to avoid amplification errors. The amplified target region can be generated by PCR on cDNA or genomic DNA (*see* **Note 2**).
7. Check the PCR results on a 1% agarose gel to ensure that a PCR product is specifically amplified at the expected size. It might be necessary to purify the PCR product.
8. Clone each PCR product in a TA vector by following the manufacturer's recommendation.

9. Transform *E. coli*-competent cells with the ligation product (*see* **Note 3**). Allow frozen competent *E. coli* cells to thaw slowly on ice.
10. Add 200 μL of the cells to ligation mix and incubate on ice for 30 min.
11. Heat-shock cells for 45 s at 42°C and chill them for 5 min on ice.
12. Add 1 mL of SOC broth and incubate for 1 h at 37°C with shaking.
13. Plate 200 μL of cells on LB agar plates containing 50 μg/mL ampicillin. Centrifuge the remaining cells and remove most of the supernatant (leave 100–200 μL in the tube). Resuspend cells in the remaining supernatant and plate them on LB agar plates with ampicillin (*see* **Note 4**). Incubate the plates at 37°C overnight.
14. Inoculate colonies in 5 mL LB broth containing 50 μg/mL ampicillin and grow the cells overnight at 37°C with shaking.
15. Prepare plasmid DNA minipreps and check the presence of the insertion by restriction digestion and agarose gel electrophoresis.
16. Confirm the absence of error in the cloned PCR fragments A1 and A2 by sequencing.

3.2. Construction of the RNAi Vectors for Plant Transformation

All steps of the cloning procedure are graphically summarized in **Fig. 4.3**:

1. For insertion of the right arm (sense), digest the TA vector containing the PCR product A1 (TA vector-A1) with *Hin*dIII and *Sal*I.
2. Open the vector pBSint11 with *Hin*dIII and *Xho*I.
3. Run the entire digested samples on 1% agarose gel.
4. Cut out the bands of interest with a clean razor blade and purify the appropriate DNA fragments using gel extraction kit.
5. Confirm quantity and quality of purified DNA fragments by analytical agarose gel electrophoresis. Fragments can be stored at –20°C for later use.
6. Ligate 30–75 ng of open vector pBSint11 with approximately threefold molar excess of the sense fragment with T4 DNA ligase in 1× ligation buffer and a final volume of 20 μL overnight at 15°C.
7. Transform *E. coli*-competent cells as described in **Section 3.1**.

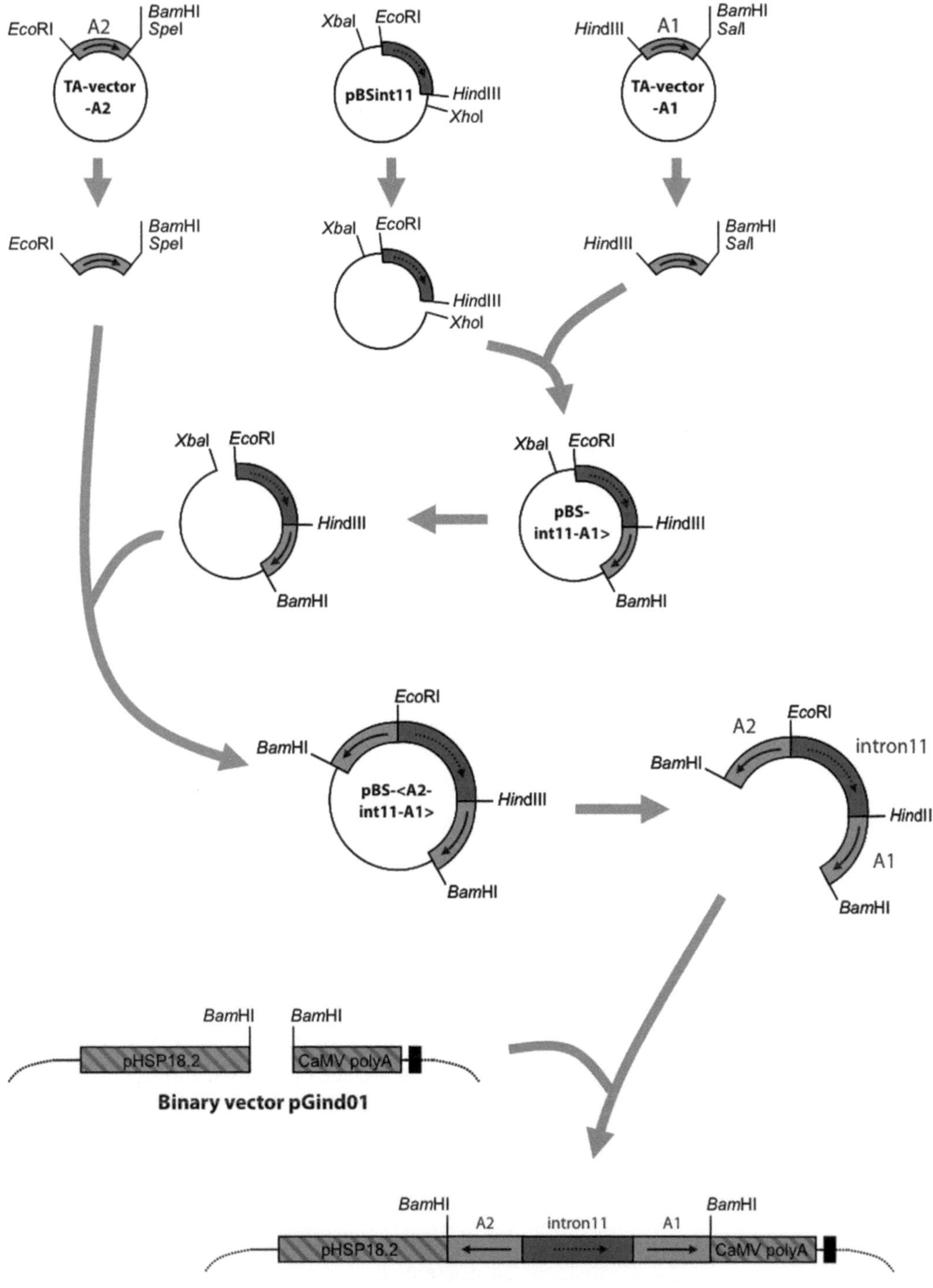

Fig. 4.3. Cloning map for generation of the constructs. This figure illustrates the cloning strategy described in the methods in **Section 3.2**.

8. Prepare plasmid DNA and control the colonies as described in **Section 3.1**.
9. Select derivatives of plasmid pBSint11 containing the sense fragment (pBS-int11-A1>) by testing with restriction enzymes and agarose gel electrophoresis.
10. For insertion of the left arm (antisense), digest the TA vector containing the PCR product A2 (TA vector-A2) with *Eco*RI and *Spe*I.
11. Open the vector pBS-int11-A1> with *Eco*RI and *Xba*I.
12. Follow the steps 3–8 described in this section.
13. Select derivatives of plasmid pBS-int11-A1> containing the antisense fragment (pBS-<A2-int11-A1>).
14. For the preparation of the binary vectors, digest the vector pGind01 containing the heat-inducible promoter with *Bam*HI (*see* **Note 5**). Do the same for the vector pSR01 containing the constitutive 35S promoter.
15. Place each mix on ice and confirm completion of the digestion by running a small aliquot of the reaction on an agarose gel.
16. Purify the open vectors with a DNA cleanup kit (e.g. Qiagen Qiaex II).
17. Dephosphorylate cleaved vectors pGind01 and pSR01 with shrimp alkaline phosphatase for 30 min at 37°C.
18. Purify immediately the DNA fragments using DNA cleanup kit. Fragments can be stored at –20°C for later use.
19. For cloning into binary vectors, digest the vector pBS-<A2-int11-A1> with *Bam*HI.
20. Run the entire digested sample on 1% agarose gel.
21. Cut out the band of interest with a razor blade and purify the appropriate DNA fragments using gel extraction kit.
22. Confirm quantity and quality of purified DNA fragments by analytical agarose gel electrophoresis. Fragments can be stored at –20°C for later use.
23. Ligate 30–75 ng of each dephosphorylated open vector with approximately threefold molar excess of the hairpin construct with T4 DNA ligase in 1× ligation buffer and a final volume of 20 μL overnight at 15°C.
24. Transform competent *E. coli* cells as described in **Section 3.1**. Use LB agar plates containing 50 μg/mL kanamycin for the selection.
25. Prepare plasmid DNA and control the colonies as previously described in **Section 3.1**.

26. Check the orientation of the construct in the vectors to select vectors with the correct orientation of the intron (*see* **Note 6**).
27. Prepare plasmid DNA and control the colonies as described in this section. Restriction digestion with the enzyme *Eco*RI produces a large band if the intron is in the correct orientation and a small band if the intron is in the reverse orientation. Select the final vectors pSR01-< A2-int11-A1> and pGind01-< A2-int11-A1>.

3.3. Transformation of Agrobacterium with the RNAi Vector

1. Inoculate a single colony of *A. tumefaciens* strain GV3101 in 5 mL YEP medium supplemented with 20 μg/mL gentamicin and 50 μg/mL rifampicin. Grow overnight at 28°C with constant shaking.
2. Inoculate 50 mL of YEP medium containing antibiotics in a 250-mL flask with 2 mL of overnight culture. Incubate at 28°C with shaking until an OD_{600} of 0.5–1 is obtained (which usually takes 7 h).
3. Cool cells on ice for 10 min and centrifuge them for 5 min at 3,000×*g* at 4°C. Remove the supernatant carefully and resuspend the cells gently in 1 mL of ice-cold 20 mM $CaCl_2$ supplemented with 15% glycerol. Competent cells can be frozen as 100 μL aliquots in liquid nitrogen at this stage and stored in a freezer at –80°C.
4. For transformation of *Agrobacterium* cells, allow 100 μL aliquots frozen cells to defreeze on ice.
5. Add 1 μg of the binary vector carrying the RNAi construct and 1 μg of the vector pSOUP and incubate for 5 min on ice.
6. Freeze cells in liquid nitrogen for 1 min and thaw the frozen cells in a water bath at 37°C for 5 min.
7. Add 1 mL of YEP medium and incubate at 28°C with gentle shaking for 4 h.
8. Centrifuge briefly at 8,000×*g* and resuspend the pellet in 100 μL YEP medium.
9. Plate the cells on YEP agar plates containing 20 μg/mL gentamicin, 50 μg/mL rifampicin, and 50 μg/mL kanamycin (*see* **Note 7**).
10. Incubate the plates at 28°C until apparition of colonies (which takes 2–3 days).
11. For the selection of the colonies containing the construct, grow individually 10 single colonies in 5 mL YEP medium supplemented with 20 μg/mL gentamicin, 50 μg/mL

rifampicin, and 50 μg/mL kanamycin. Incubate overnight at 28°C with constant shaking.

12. Keep a backup of the inoculated colonies on a YEP agar plate with the appropriate antibiotics.
13. Prepare plasmid DNA minipreps from the overnight cultures and perform restriction digestion and/or PCR to confirm that the cells contain the right construct.

3.4. Transformation of Arabidopsis Plants

1. Place the *Arabidopsis* seeds in pots at low density (to get not more than three plants in an 8-cm × 8-cm square pot, *see* **Note 8**).
2. After 2 days of stratification at 4°C, grow the plants in short days (9–12 h of light per day) with fertilizer, to obtain large rosettes (*see* **Note 9**). To help hold the soil in the pot, it is possible to cover the pots with nylon mesh.
3. After 3–4 weeks in short days, transfer the plants to long days (16 h of light per day) to trigger flowering.
4. When the plants start to flower, remove the emerging primary inflorescence to encourage growth of multiple new inflorescences.
5. Plants are optimal to be transformed when there are a large number of unopened flower buds (*see* **Note 10**).
6. Before plants transformation, inoculate a single colony of *A. tumefaciens* containing the RNAi construct in 10 mL YEP medium supplemented with 20 μg/mL gentamicin, 50 μg/mL rifampicin, and 50 μg/mL kanamycin. Grow overnight at 28°C with constant shaking.
7. Inoculate 500 mL of YEP medium containing antibiotics in a 2-L flask with 5 mL of overnight culture. Incubate overnight at 28°C with shaking until an OD_{600} of 0.8–1 is obtained.
8. Centrifuge the cells for 10 min at 2,500×*g* at room temperature. Resuspend the pellet in 160 mL of transformation medium (one-third of the culture volume).
9. Transfer the cell suspension in a beaker.
10. Invert a pot containing the plants to be transformed (named T0) and submerge the inflorescences into the cell suspension. Shake slowly for 30 s. Repeat this step with all the pots.
11. Put the plants in a tray and cover them with Saran wrap to maintain high humidity (*see* **Note 11**). Transfer the tray in the dark at growth temperature (22°C) for 24 h. Remove the Saran wrap and grow the plant normally until the seeds are ready to be harvested.

12. Harvest the seeds (T1) individually for each pot and let them dry for 1 week at room temperature.

3.5. Selection of the RNAi Transgenic Plants

1. Put 100–200 μL of seeds in a microtube. Sterilize the seeds by incubating them not more than 10 min in the sterilization solution. Shake the tube three times during the 10 min. Working in a sterile hood, remove the solution and rinse the seeds three times with 95% ethanol. Transfer the seeds to sterile-filter papers and let them dry for 15 min.
2. Sow the seeds on the surface of agar plates containing half-strength MS medium supplemented with 50 μg/mL kanamycin. Seal the plates with microporous tape and incubate at 4°C for 2 days. Transfer the plate to a standard plant growth room.
3. Identify the resistant, green, and healthy plants among the non-transformed plants around 10 days after transfer to the growth room. Transfer the resistant plants to soil. Maintain high humidity with a cover during 2 days. Grow the plant until collecting individually the T2 seeds for each plant (*see* **Note 12**).
4. Sow sterilized T2 seeds on agar plates containing half-strength MS medium supplemented with 50 mg/mL kanamycin to select only the plants harbouring the RNAi construct. Submit these transgenic plants directly to the heat treatment or grow them in soil prior to the heat treatment (*see* **Note 13**).

3.6. Heat Induction of the RNAi Construct and Plant Characterization

1. For heat induction of in vitro cultured plants, grow the seedlings in agar plates sealed with microporous tape in a standard growth room. Avoid all kinds of stresses which could induce the *HSP18.2* promoter (*see* **Note 14**). Include WT control plants.
2. For the induction of the RNAi cassette expression, incubate the plates for 2 h in an incubator at 37°C. Then return the plates to the growth room.
3. For heat induction of plants cultured in pots, grow the plants on soil in a standard growth room. Be sure to prevent these plants from all kinds of stresses which could induce the *HSP18.2* promoter, especially drought (*see* **Note 14**). Include WT control plants.
4. Prepare the heat-inducing chamber: add prewarmed water (37°C) and place it in an incubator regulated at 37°C (*see* **Note 15**)
5. Transfer the plants in the incubator, close carefully the chamber, and start the fog generator.

6. Keep the plants in the incubator for 30–45 min.
7. Return the plants to the growth room.
8. For in vitro cultured plants or plants cultured in soil-filled pots, to maintain persistent downregulation of the gene of interest, repeat the heat induction step every 2 days (*see* **Note 16**).
9. After 2–3 days (*see* **Note 17**), observe the plants for seedlings for phenotypic alterations or collect them for molecular analysis.
10. For gene expression analysis, prepare total RNA from WT, induced RNAi transgenic plants, and non-induced RNAi transgenic plants. Perform quantitative RT-PCR or northern blot to quantify the expression of the target gene in wild-type control and induced RNAi transgenic lines by referring to standard methods. The target gene should be downregulated in the induced RNAi transgenic lines compared to WT and non-induced transgenic lines.
11. Phenotype the induced RNAi plants in the frame of your scientific context.

4. Notes

1. Cross-homology silencing can appear if there is a block of contiguous 19–20 bases identical between the construct and non-target gene sequences. It is important to check if such blocks do not exist in the complete genome (32, 33).
2. The choice of the template for the amplification is an important parameter. If the target region corresponds to a genomic region without intron, genomic DNA can be used for the amplification. If cDNA is used as template, it is necessary to prepare cDNA from plant samples (organ, conditions) in which the gene of interest is expressed.
3. We routinely use the Inoue's method to obtain highly competent *E. coli* cells (10^7–10^8 cfu/μg DNA). With this method, cells are grown at low temperature (20°C) and DMSO is used to permeabilize cell membranes (34).
4. Cell spreading can be achieved quickly and efficiently with glass beads instead of glass hook. The use of glass beads minimizes the risk of contamination. Use 4-mm glass beads sterilized by autoclaving. Put around five beads per plate prior to adding the bacteria. Once the bacteria solution is added to the plate, shake the plate horizontally in all direction for 30 s to 1 min. Return the plates and remove

the beads. Glass beads can be washed and reused after autoclaving.

5. If another cloning strategy is used, the *HSP18.2* promoter (*pHSP18.2*) can be amplified following instruction of a previous work (23).
6. To ensure efficient silencing, the intron has to be spliced off the construct which will then form a hairpin structure. Only an intron in the correct orientation will be spliceable.
7. The pSOUP helper vector which confers resistance to tetracycline (5 μg/mL for *E. coli*) is necessary to maintain the pGreen-derivative vectors pSR01 and pGind01 in *Agrobacterium*. As a consequence, it is not necessary to add tetracycline in the medium. Furthermore, it is recommended to always co-transform pSOUP and pGreen vectors in *Agrobacterium* as pSOUP alone was reported to be not completely stable in the bacteria.
8. As healthy plants are easier to transform than stressed plants, do not grow too much plants/pot as competition will be a stressing factor.
9. Use fertilizer several times to produce very healthy big plants. If time is limited, grow directly the plants under long day conditions.
10. As a minimum of 15 transgenic plants are required for the subsequent analysis, it is necessary to transform at least 20 plants.
11. If possible, it is really more convenient to use a plastic tray with a plastic cover. The Saran wrap can sometimes touch the plants and it is difficult to remove it without damaging the plants. The objective being to maintain high humidity atmosphere, a plastic cover is really easier to manipulate than is Saran wrap.
12. The RNAi construct is functional in the T1 plants. It is possible to use some of these plants to get preliminary data.
13. Alternatively, you can select mono-insertion homozygous lines in T3 generation by segregation analysis, PCR, and Southern Blot. In this case, no selection is required when working with these isolated lines. Furthermore, the efficiency will be very reproducible. Nevertheless, it is necessary to check several different lines to ensure the RNAi efficiency.
14. The *HSP18.2* promoter can be induced by several stress conditions (e.g. heat, nematodes, salt, high light, and hypoxia). Grow the plant in very controlled conditions with a regular limited watering to ensure the inactivity of the *HSP18.2* promoter.

15. An efficient induction requires a temperature of 37°C in a very wet atmosphere. Alternative chambers should conform to these criteria.
16. The induction of the *HSP18.2* promoter is transient after heat treatment. The silencing of the gene of interest via the RNAi construct is also transient. A repeated induction might be necessary to maintain a constant level of the RNAi construct expression as described (23).
17. From the heat induction to the effective downregulation of the target gene, a significant period of time is required. We observed that the complete downregulation of a target gene takes around 24 h. Furthermore, when targeting the *PDS* gene, the resulting photobleaching appears only 2–3 days after induction. The time necessary for the downregulation of the target gene and the appearance of a phenotype should be specific for each target gene.

Acknowledgements

We thank Phil Mullineaux and Roger Hellens from John Innes Centre and the BBSRC for providing us with *pGREEN* and *pSOUP* vectors. We are grateful to Rafaël Pont-Lezica, Séverine Lorrain, and Martine Charpenteau for discussion and critical reading of the manuscript.

References

1. Arabidopsis Genome Initiative (2000) Analysis of the genome sequence of the flowering plant *Arabidopsis thaliana*. *Nature* **408**, 796–815.
2. Fox, S., Flichkin, S., and Mockler, T. C. (2009) Applications of ultra-high-throughput sequencing. In: *Plant Systems Biology* (ed. Belostotsky, D.) *Methods in Molecular Biology*, Humana Press, Totowa, NJ, **Vol. 553**, pp. 79–108.
3. Bourque, J. E. (1995) Antisense strategies for genetic manipulations in plants. *Plant Sci.* **105**, 125–149.
4. Chuang, C. F. and Meyerowitz, E. M. (2000) Specific and heritable genetic interference by double-stranded RNA in *Arabidopsis thaliana*. *Proc. Natl. Acad. Sci. USA* **97**, 4985–4990.
5. Tavernarakis, N., Wang, S. L., Dorovkov, M., Ryazanov, A., and Driscoll, A. M. (2000) Heritable and inducible genetic interference by double-stranded RNA encoded by transgenes. *Nat. Genet.* **24**, 180–183.
6. Sui, G., Soohoo, C., Affar, B., Gay, F., Shi, Y., Forrester, W. C., and Shi, Y. (2002) A DNA vector-based RNAi technology to suppress gene expression in mammalian cells. *Proc. Natl. Acad. Sci. USA* **99**, 5515–5520.
7. Fire, A., Xu, S., Montgomery, M. K., Kostas, S. A., Driver, D. E., and Mello, C. C. (1998) Potent and specific genetic interference by double-stranded RNA in *Caenorhabditis elegans*. *Nature* **391**, 806–811.
8. Waterhouse, P. M., Graham, M. W., and Wang, M. B. (1998) Virus resistance and gene silencing in plants can be induced by simultaneous expression of sense and antisense RNA. *Proc. Natl. Acad. Sci. USA* **95**, 13959–13964.

9. Bernstein, E., Caudy, A. A., Hammond, S. M., and Hannon, G. J. (2001) Role for a bidentate ribonuclease in the initiation step of RNA interference. *Nature* **409**, 363–366.
10. Cigan, A. M., Unger-Wallace, E., and Haug-Collet, K. (2005) Transcriptional gene silencing as a tool for uncovering gene function in maize. *Plant J.* **6**, 929–940.
11. Mansoor, S., Amin, I., Hussain, M., Zafar, Y., and Briddon, R. W. (2006) Engineering novel traits in plants through RNA interference. *Trends Plant Sci.* **11**, 559–565.
12. Schwab, R., Ossowski, S., Riester, M., Warthmann, N., and Weigel, D. (2006) Highly specific gene silencing by artificial microRNAs in *Arabidopsis. Plant Cell* **18**, 1121–1133.
13. Yang, Y., Costa, A., Leonhardt, N., Siegel, R. S., and Schroeder, J. I. (2008) Isolation of a strong *Arabidopsis* guard cell promoter and its potential as a research tool. *Plant Methods* **4**, 1–15.
14. Gatz, C. and Lenk, I. (1998) Promoters that respond to chemical inducers. *Trends Plant Sci.* **3**, 352–358.
15. Gatz, C., Frohberg, C., and Wendenburg, R. (1992) Stringent repression and homogeneous derepression by tetracycline of a modified CaMV 35S promoter in intact transgenic tobacco plants. *Plant J.* **2**, 397–404.
16. Mett, V. L., Lochhead, L. P., and Reynolds, P. H. (1993) Copper-controllable gene expression system for whole plants. *Proc. Natl. Acad. Sci. USA* **90**, 4567–4571.
17. Ait-Ali, T., Rands, C., and Harberd, N. P. (2003) Flexible control of plant architecture and yield via switchable expression of *Arabidopsis gai. Plant Biotechnol. J.* **1**, 337–343.
18. Zuo, J. and Chua, N. H. (2000) Chemical-inducible systems for regulated expression of plant genes. *Curr. Opin. Biotechnol.* **11**, 146–151.
19. Andersen, S. U., Cvitanich, C., Hougaard, B. K., Roussis, A., Gronlund, M., Jensen, D. B., Frokjaer, L. A., and Jensen, E. O. (2003) The glucocorticoid-inducible GVG system causes severe growth defects in both root and shoot of the model legume *Lotus japonicus. Mol. Plant Micr. Interact.* **16**, 1069–1076.
20. Vreugdenhil, D., Claassens, M. M., Verhees, J., van der Krol, A. R., and van der Plas, L. H. (2006) Ethanol-inducible gene expression: non-transformed plants also respond to ethanol. *Trends Plant Sci.* **11**, 9–11.
21. Takahashi, T. and Komeda, Y. (1989) Characterization of two genes encoding small heat-shock proteins in *Arabidopsis thaliana. Mol. Gen. Genet.* **219**, 365–372.
22. Takahashi, T., Naito, S., and Komeda, Y. (1992) The *Arabidopsis HSP18.2* promoter/*GUS* gene fusion in transgenic *Arabidopsis* plants: a powerful tool for the isolation of regulatory mutants of the heat-shock response. *Plant J.* **2**, 751–761.
23. Matsuhara, S., Jingu, F., Takahashi, T., and Komeda, Y. (2000) Heat-shock tagging: a simple method for expression and isolation of plant genome DNA flanked by T-DNA insertions. *Plant J.* **22**, 79–86.
24. Yoshida, K. and Shinmyo, A. (2000) Transgene expression systems in plant, a natural bioreactor. *J Biosci. Bioeng.* **90**, 353–362.
25. Luo, K., Sun, M., Deng, W., and Xu, S. (2008) Excision of selectable marker gene from transgenic tobacco using the GM-gene-deletor system regulated by a heat-inducible promoter. *Biotechnol. Lett.* **30**, 1295–1302.
26. Masclaux, F., Charpenteau, M., Takahashi, T., Pont-Lezica, R., and Galaud, J. P. (2004) Gene silencing using a heat-inducible RNAi system in *Arabidopsis. Biochem. Biophys. Res. Commun.* **321**, 364–369.
27. Guo, H. S., Fei, J. F., Xie, Q., and Chua, N. H. (2003) A chemical-regulated inducible RNAi system in plants. *Plant J.* **34**, 383–392.
28. Ruiz, M. T., Voinnet, O., and Baulcombe, D. C. (1998) Initiation and maintenance of virus-induced gene silencing. *Plant Cell* **10**, 937–946.
29. Wesley, S. V., Helliwell, C. A., Smith, N. A., Wang, M. B., Rouse, D. T., Liu, Q., Gooding, P. S., Singh, S. P., Abbott, D., Stoutjesdijk, P. A., Robinson, S. P., Gleave, A. P., Green, A. G., and Waterhouse, P. M. (2001) Construct design for efficient, effective and high-throughput gene silencing in plants. *Plant J.* **27**, 581–590.
30. Hellens, R. P., Edwards, E. A., Leyland, N. R., Bean, S., and Mullineaux, P. M. (2000) pGreen: a versatile and flexible binary Ti vector for *Agrobacterium*-mediated plant transformation. *Plant Mol. Biol.* **42**, 819–832. www.pgreen.ac.uk
31. Koncz, C. and Schnell, J. (1986) The promoter of TL-DNA gene *5* controls the tissue specific expression of chimeric genes carried by a novel type of *Agrobacterium* binary vector. *Mol. Gen. Genet.* **204**, 383–396.
32. Kulkarni, M. M., Booker, M., Silver, S. J., Friedman, A., Hong, P., Perrimon, N., and Mathey-Prevot, B. (2006) Evidence of off-target effects associated with long dsRNAs in *Drosophila melanogaster* cell-based assays. *Nat. Methods* **10**, 833–838.
33. Helliwell, C. and Waterhouse, P. (2003) Constructs and methods for high-throughput gene silencing in plants. *Methods* **30**, 289–295.
34. Inoue, H., Nojima, H., and Okayama, H. (1990) High efficiency transformation of *Escherichia coli* with plasmids. *Gene* **96**, 23–28.

Chapter 5

Gene Function Analysis by Artificial MicroRNAs in *Physcomitrella patens*

Basel Khraiwesh, Isam Fattash, M. Asif Arif, and Wolfgang Frank

Abstract

MicroRNAs (miRNAs) are ~21 nt long small RNAs transcribed from endogenous *MIR* genes which form precursor RNAs with a characteristic hairpin structure. miRNAs control the expression of cognate target genes by binding to reverse complementary sequences resulting in cleavage or translational inhibition of the target RNA. Artificial miRNAs (amiRNAs) can be generated by exchanging the miRNA/miRNA* sequence of endogenous *MIR* precursor genes, while maintaining the general pattern of matches and mismatches in the foldback. Thus, for functional gene analysis amiRNAs can be designed to target any gene of interest. During the last decade the moss *Physcomitrella patens* emerged as a model plant for functional gene analysis based on its unique ability to integrate DNA into the nuclear genome by homologous recombination which allows for the generation of targeted gene knockout mutants. In addition to this, we developed a protocol to express amiRNAs in *P. patens* that has particular advantages over the generation of knockout mutants and might be used to speed up reverse genetics approaches in this model species.

Key words: Artificial microRNA, gene knockdown, microRNA, *Physcomitrella patens*, RNA interference.

1. Introduction

Small RNAs, 20–24 nucleotides (nt) long non-protein coding RNAs, have been increasingly investigated. They were responsible for phenomena described as RNA interference (RNAi), co-suppression, gene silencing, or quelling (1–4). Since the discovery of RNAi in *Caenorhabditis elegans* (2), extensive studies have been performed focusing on the different aspects of RNAi. In particular, the elucidation of the essential components of RNAi pathways has extensively advanced (5). Shortly after the discovery of RNAi,

H. Kodama, A. Komamine (eds.), *RNAi and Plant Gene Function Analysis*, Methods in Molecular Biology 744, DOI 10.1007/978-1-61779-123-9_5, © Springer Science+Business Media, LLC 2011

it was shown that posttranscriptional gene silencing (PTGS) in plants is correlated with the presence of a population of small RNAs and contains both sense and antisense RNA sequences (6).

Major classes of small RNAs include microRNAs (miRNAs) and small interfering RNAs (siRNAs), which differ in their biosynthesis (7, 8). miRNAs are ~21 nt long small RNAs transcribed from endogenous *MIR* genes forming precursor RNAs with a characteristic hairpin structure that are generated from longer pre-miRNA precursors by Dicer-like (DCL) proteins in plants. These miRNAs are recruited to the RNA-induced silencing complex (RISC) (7, 9). Recently, miRNAs have been identified as important regulators of gene expression in both plants and animals (10), and they are highly conserved in evolution (10–12). siRNAs have similar structure and function as miRNAs, but they derive from long double-stranded RNA (dsRNA) of transgenes, endogenous repeat sequences, or transposons (13, 14). *Trans-acting* small interfering RNAs (ta-siRNA), like miRNAs, are originated from nuclear-encoded *TAS* loci which give rise to non-coding transcripts. Processing of *TAS* precursors requires miRNA-directed cleavage of the transcripts, subsequent conversion into dsRNA by RNA-dependent RNA polymerases, and phased slicing of the dsRNA by DCL proteins into distinct ta-siRNAs (8, 15–19). Some of the generated ta-siRNAs act like miRNAs by binding to cognate target mRNAs to control their expression at the posttranscriptional level (11, 15, 20, 21). The mode of action of small RNAs to control gene expression at the posttranscriptional level is now being developed into tools for biological research. In principle, the design of small RNA molecules that are able to target any RNA of interest in a highly specific manner allows the precise knockdown of RNAs for gene function analyses. Previous studies have shown that the alteration of several nucleotides within the miRNA/miRNA* sequence (miRNA*: complementary sequence of the mature miRNA) of a *MIR* precursor does not affect its biogenesis and maturation as long as the number of matches and mismatches in the precursor backbone, and thus the structure, are maintained (22, 23). This finding raises the possibility to modify miRNA sequences within *MIR* precursors according to the sequence of a selected target RNA. Thus, miRNAs can be generated artificially that specifically target any intended RNA. This technology of artificial microRNA (amiRNA) expression successfully exploits endogenous miRNA precursors to generate small RNAs that direct gene silencing in either plants or animals (24–29). Furthermore, amiRNA sequences can be easily optimized to silence one or several target transcripts without affecting the expression of other transcripts (off-targets). In humans the miR30 precursor has been modified to generate an amiRNA that downregulates target gene expression by translation inhibition (30, 31). *Arabidopsis thaliana* miRNA precursors have been modified to silence endogenous as well as exogenous target genes in the dicotyledonous plants *A. thaliana*, tomato, and tobacco (24–27, 32). Gene silencing by amiRNAs in the monocotyledonous species *Oryza sativa* has been reported as well (28), and recently we established an amiRNA expression system for the moss *Physcomitrella patens* (33). Gene knockdown approaches using conventional RNAi constructs rely on the expression of considerably long inverted stretches of DNA. In consequence, the expression of these constructs results in the generation of a diverse set of siRNAs that increases the possibility to affect unintended off-targets. In contrast, amiRNAs can be designed either to target a specific RNA in a highly specific manner or to target a group of sequence-related homologous RNAs. In addition, chemically induced promoters as well as tissue-specific promoters can be used to drive amiRNA expression to achieve a controlled downregulation of the intended target RNA (24, 27). In this chapter we provide protocols for the (i) design and engineering of amiRNAs, (ii) the generation of amiRNA-expressing *P. patens lines* and the analysis of these lines including (iii) the analysis of amiRNA expression levels, (iv) the analysis of amiRNA target transcripts, and (v) the detection of amiRNA target cleavage products.

2. Materials

2.1. Culture Media and Growth Conditions for P. patens

1. Growth medium: 250 mg/L KH_2PO_4, 250 mg/L KCl, 250 mg/L $MgSO_4$, 1,000 mg/L $Ca(NO_3)_2$, and 12.5 mg/L $FeSO_4$, pH 5.8. Solid medium contains 12 g/L agar. Sterilize by autoclaving.
2. Growth medium for protoplast isolation (reduced $Ca(NO_3)_2$): 250 mg/L KH_2PO_4, 250 mg/L KCl, 250 mg/L $MgSO_4$, 100 mg/L $Ca(NO_3)_2$, and 12.5 mg/L $FeSO_4$, pH 5.8. Solid medium contains 12 g/L agar. Sterilize by autoclaving.
3. Erlenmeyer flasks containing 400 mL of suspension culture agitated on a rotary shaker at 120 rpm at 25°C under a light/dark regime of 16/8 h (Philips TLD 25, 50 μM m^{-2} s^{-1}).

2.2. Protoplasts Isolation, Transfection, and Regeneration

1. Driselase 2% (w/v) (Sigma Chemical Co., Dorset, UK) in 0.5 M mannitol, filter sterilize using a 0.22 μm filter.
2. Sieves (Wilson, UK) of 100 μm and 45 μm pore sizes.
3. 3M medium: 5 mM $MgCl_2$, 0.1% (w/v) 2-(*N*-morpholino) ethanesulfonic acid (MES), 0.48 M mannitol, pH 5.6. The

osmolarity of the solution should be approximately 560 mOs. Sterilize by autoclaving.

2.3. Transformation of P. patens Protoplasts

1. Polyethylene glycol (PEG): 40% (w/v) PEG_{4000} in 3M medium, pH 6.0. Filter sterilize using a 0.22 μm filter.
2. For each transformation we need 25 μg linearized plasmid DNA including vector backbone or 10–15 μg of linearized vector-free plasmid DNA. Purify the DNA by standard ethanol precipitation and dissolve the construct at a final concentration of 0.25 μg/μL in 100 μL of 0.1 M $Ca(NO_3)_2$ (dissolve the DNA pellet in a clean bench to keep the DNA sample sterile for the protoplast transformation).
3. Autoclaved cellophane sheets.
4. Selection medium: Prepare standard solid growth medium and autoclave. Cool the medium to 60°C and add Hygromycin-B (Sigma-Aldrich) to reach the desired final antibiotic concentration (12.5 μg/mL). Pour the medium into petri dishes. The plates can be stored up to 4 weeks at 4°C. Hygromycin-B is used when the expression vector harbors an *hpt* expression cassette. The antibiotic depends on the selection marker cassette that is present in the expression vector of choice.
5. Regeneration medium: For 1 L use 10 mL of each stock solution described for preparation of growth medium (KH_2PO_4, KCl, $MgSO_4$, $Ca(NO_3)_2$), add 12.5 mg $FeSO_4$, add 50 g glucose, add 30 g mannitol, and adjust pH to 5.8 with KOH. Adjust the osmolarity to approximately 540 mOs using mannitol. Filter sterilize through a 0.22 μm filter.

2.4. One-Step Isolation of Genomic DNA from P. patens

1. 10X DNA extraction buffer: To prepare 1 L dissolve 90.86 g Tris, 26.43 g $(NH_4)_2SO_4$, 1 mL Tween-20 in 800 mL of H_2O, adjust pH to 8.8 with HCl; before use dilute 1:10 with H_2O.
2. 96-well PCR plate for parallel isolation of genomic DNA from 96 transformants, 96-well thermocycler.

2.5. PCR-Based Analysis

1. 10X PCR buffer with $(NH_4)_2SO_4$: 750 mM Tris–HCl, pH 8.8, 200 mM $(NH_4)_2SO_4$, 0.1% Tween-20 (Promega), 25 mM $MgCl_2$ (Promega), 2 mM dNTP mix (MBI Fermentas), Taq DNA Polymerase (Promega).
2. 3 mM spermidine solution: Dissolve 38.2 mg spermidine trihydrochloride (Sigma, S2501) in 50 mL of H_2O, sterilize by filtration through a 0.22 μm filter.
3. PCR primers.
4. Thermocycler.

2.6. Isolation of Total RNA from *P. patens* Using TRIzol Reagent

1. TRIzol reagent (Invitrogen).
2. Chloroform.
3. Isopropanol.
4. 75% ethanol in DEPC-treated H_2O.
5. RNase-free water (GIBCO).

2.7. Real-Time PCR (RT-PCR)

1. TaqMan Kit (Applied Biosystems).
2. *PpEF1* primer pairs: forward primer 5′-CGACGCCCCTGG ACATC-3′; reverse primer 5′-CCTGCGAGGTTCCCG TAA-3′.
3. SensiMix dT Kit (Quantace).

2.8. Small RNA Northern Blotting (Detection and Analysis of amiRNAs)

1. 10X TBE buffer: 900 mM Tris–borate, 20 mM EDTA.
2. Polyacrylamide gel: Rotiphorese Gel 40 (37.5:1) (Roth), 10X TBE buffer, 8.M urea, 10% ammonium persulfate (APS), *N,N,N′,N′*-tetramethylethylenediamine (TEMED, pH 7.0).
3. Electrophoresis buffer: 0.5X TBE.
4. 2X RNA loading buffer: 98% deionized formamide, 2 mM EDTA, 0.01% xylene cyanol, 0.01% bromophenol blue.
5. Decade RNA Marker (Ambion), stored at −20°C.
6. Hybond N^+ nylon membrane (GE Healthcare).
7. Whatman paper: GB002 grade.
8. Electroblotter: Trans-Blot Semi-Dry Electrophoretic Transfer Cell (Bio-Rad).
9. UV crosslinker (Biolink BLX, Biometra).
10. DNA oligonucleotides complementary to amiRNA.
11. 10 μCi/μL [^{32}P]-ATP: ~6,000 Ci/mmol.
12. 10 U/μL T4 polynucleotide kinase (Fermentas).
13. Nucleotide Removal Kit (Qiagen).
14. 20X SSC: 3M NaCl, 0.3M sodium acctatc, pII 7.0.
15. Hybridization buffer: 0.05 M sodium phosphate, pH 7.2, 1 mM EDTA, 6X SSC, 1X Denhardt's solution, 5% SDS.
16. Washing solutions: 2X SSC, 0.2% SDS and 1X SSC, 0.1% SDS.
17. PhosphorImager (Imager FX, Bio-Rad).

2.9. Northern Blotting (Analysis of amiRNA Target Transcripts)

1. 10X MOPS: 0.2 M MOPS, 20 mM sodium acetate, 10 mM EDTA, DEPC-H_2O, pH 7.0.
2. 20 μg of total RNA.

3. RNA denaturing buffer: 500 μL deionized formamide, 120 μL 37% (w/v) formaldehyde, 200 μL 10X MOPS, and 1 μL ethidium bromide.
4. 10X RNA loading buffer: 50% glycerol, 0.25% bromophenol blue, DEPC-H_2O (or RNase-free H_2O, e.g., from GIBCO).
5. Agarose formaldehyde gel: 1.5% agarose, 1X MOPS, 6.5% formaldehyde.
6. RNA size marker, e.g., peqGOLD High Range RNA Ladder (PeqLab).
7. Electrophoresis buffer: 1X MOPS (without formaldehyde).
8. Hybond N^+ nylon membrane (GE Healthcare).
9. Whatman paper: GB002 grade and GB004 grade.
10. Turboblotter (Schleicher & Schuell) with 20X SSC.
11. UV crosslinker (Biolink BLX, Biometra).
12. Hybridization buffer: 1 M Na_2HPO_4, 1 M NaH_2PO_4, 0.5 M EDTA, pH 8.0, 20% SDS, 100 μg/mL denatured salmon sperm DNA.
13. TE buffer: 10 mM Tris–Cl, pH 8.0, 1 mM EDTA.
14. [^{32}P]-dCTP-labeled DNA probe.
15. Rediprime II Random Prime Labeling System (GE Healthcare).
16. Washing solutions: 1X SSC, 0.1% SDS and 0.5X SSC, 0.1% SDS.
17. PhosphorImager (Imager FX, Bio-Rad).
18. Stripping buffer: 0.1% SDS heated to 100°C.

2.10. 5′ RACE PCR

1. Oligotex mRNA mini kit (Qiagen).
2. RNA adaptor: 5′-gacuggagcacgaggacacugacauggacugaagg aguagaaa-3′.
3. 10 U/μL T4 RNA ligase (Fermentas).
4. 200 U/μL SuperScript III reverse transcriptase (Invitrogen).
5. 40 U/μL RNaseOut (Invitrogen).
6. 2 U/μL RNase H (Invitrogen).
7. Oligo$(dT)_{18}$ primer.
8. Platinum Pfx DNA Polymerase (Invitrogen).
9. 5′ RACE adaptor primer: 5′-gcacgaggacacugacaugga cuga-3′.
10. 5′ RACE nested adaptor primer: 5′-ggacactgacatggactgaa ggagta-3′.

11. Phenol:chloroform:isoamyl alcohol (25:24:1).
12. 3M sodium acetate, pH 5.2.
13. 95 and 70% ethanol.
14. GlycoBlue (Applied Biosystems).
15. QIAquick gel extraction kit (Qiagen).

2.11. Electrophoretic Separation of DNA

1. 1–2% agarose gel.
2. 50X TAE buffer: 2 M Tris–HCl, pH 8.0, 1 M glacial acetic acid, 50 mM EDTA, pH 8.0.

3. Methods

3.1. Design of amiRNAs Using the WMD Platform

An amiRNA targeting the gene of interest can be designed using the amiRNA designer interface WMD (Web MicroRNA Designer; wmd3.weigelworld.org) (14, 27). This tool has been extended to >30 species for which genome or extensive EST information is available (*see* Table 1 in ref. 14). The designed amiRNA has to meet certain criteria, which have to be considered when choosing the final amiRNA (*see* **Note 1**).

1. The designed amiRNA contains an uridine residue at position 1 and an adenine residue at position 10, both of which are overrepresented among natural plant miRNAs and increase the efficiency of miRNA-mediated target cleavage (34).
2. The amiRNA exhibits 5′ instability relative to the miRNA* which positively affects separation of both strands during RSIC loading (34, 35).
3. No mismatch between positions 2 and 12 of the amiRNA for all targets.
4. Similar mismatch pattern for all intended targets.
5. Absolute hybridization energy between –35 and –38 kcal/mol.
6. Target site position: There is no evidence that the position of the target site in the target transcript has an effect on effectiveness, but target sites in most endogenous miRNA targets are found toward the 3′-end of the coding regions.

3.2. Engineering an amiRNA Precursor by Overlapping PCR

After selection of an amiRNA, the designed 21 nt amiRNA sequence has to be engineered into a miRNA precursor in order to replace the endogenous miRNA/miRNA* sequences within the precursor backbone. This is achieved by overlapping PCR using a miRNA precursor fragment that is cloned into a common cloning vector (27). To generate an amiRNA precursor that can be used for amiRNA expression in *P. patens* we have used the *A. thaliana* miR319a precursor backbone (*see* **Note 2**). MiR319 and the corresponding *MIR319* loci are conserved between *P. patens* and *A. thaliana* and we were able to show that amiRNAs are correctly processed from the *A. thaliana* miR319 precursor sequence in *P. patens* (33).

1. The WMD (Web MicroRNA Designer; wmd3. weigel-world.org) interface includes the "Oligo" software tool which allows automatic design of oligonucleotide primers I–IV (primer I: miRNA sense, primer II: miRNA antisense, primer III: miRNA* sense, primer IV: miRNA* antisense; **Fig. 5.1**) which are used for overlapping PCR. Overlapping PCR is carried out using the plasmid pRS300 as template which harbors the *A. thaliana MIR319a* precursor cloned into the *Sma*I site of pBSK. The oligonucleotide primers A and B are based on the plasmid vector backbone sequence. Amplify three individual PCR products using the primer

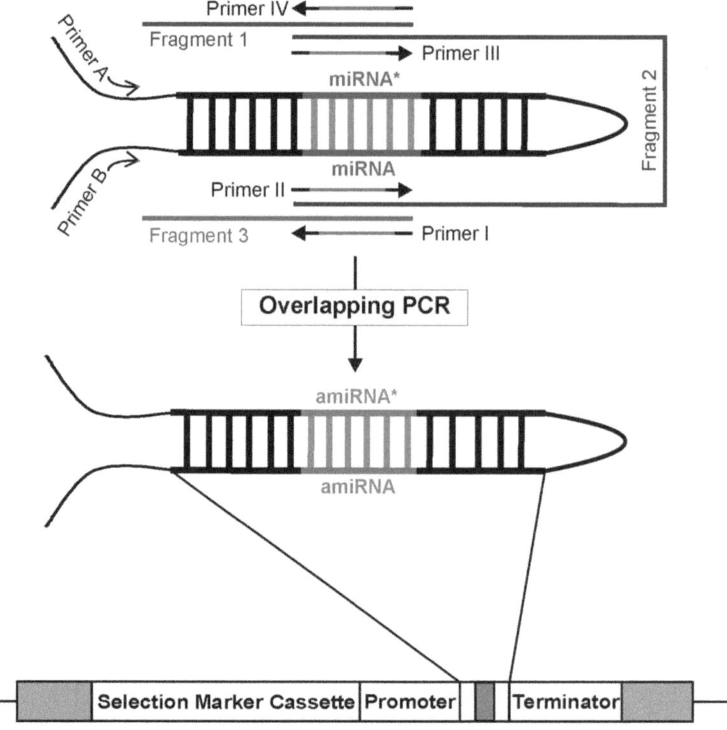

Fig. 5.1. Engineering an amiRNA sequence by overlapping PCR. Oligonucleotide primers I–IV are used to replace miRNA and miRNA* regions by amiRNA sequences. The plasmid pRS300 containing the *A. thaliana* miR319a precursor is used as PCR template. First, amplify fragment 1 with primers A and IV, fragment 2 with primers II and III, and fragment 3 with primers I and B. Primers A and B are based on template plasmid sequence. In a second round of PCR combine fragments 1, 2, and 3 as a template and perform an overlapping PCR using the primers A and B. Finally, clone the modified precursor harboring the engineered amiRNA sequence into a suitable expression vector.

pairs A and IV, II and III, and I and B, respectively. Combine the obtained PCR products and amplify the engineered amiRNA precursor in a final PCR using the primers A and B (**Fig. 5.1**).

2. The DNA fragment obtained from the overlapping PCR can be cloned into common cloning vectors. After cloning, successful exchange of the miRNA/miRNA* sequences should be confirmed by sequencing. For subcloning of the amiRNA precursor fragment into a suitable plant expression vector any sites of the pBSK multiple cloning site might be used since they are part of the PCR fragment. For the expression of amiRNAs in *P. patens* we have successfully used the vector pPCV which drives expression of the amiRNA from a CaMV *35S* promoter and harbors an *hpt* selection marker cassette (36).

3.3. Transformation of P. patens Protoplasts

PEG-mediated transformation of *P. patens* protoplasts was performed according to standard procedures (37).

1. Small-scale preparation of moss material: Starting from a standard *P. patens* liquid culture grown in Erlenmeyer flasks inoculate 200 mL of growth medium for protoplast isolation (reduced $Ca(NO_3)_2$) with moss material corresponding to 10 mg dry weight. Grow the culture in a 500 mL Erlenmeyer flask under standard conditions. After 4 days, exchange the medium and cultivate for another 3 days. The complete culture is used for protoplast isolation. Alternatively, *P. patens* plants can be grown in a 5 L bioreactor in standard growth medium at a reduced pH of 4.5. For protoplast isolation, 100–200 mL of these cultures are used.
2. Harvest the plant material by filtration through a 100 μm protoplast sieve. Transfer the material to a petri dish (9 cm diameter) and add 8 mL of 0.5 M mannitol solution. Add 8 mL of the Driselase stock solution. The final concentration of Driselase is 2% (w/v). Seal the petri dish with Parafilm and cover it with aluminum foil. Incubate for 45 min at room temperature on a rotary shaker.
3. Pass the moss material successively through sieves with a mesh size of 100 and 45 μm and divide the filtrate into two glass tubes. Centrifuge the filtrate for 10 min at 45 ×*g* in the glass tube. Discard the supernatant and wash the protoplasts by resuspending each pellet in 10 mL of 0.5 M mannitol and gentle rolling of the glass tubes between your hands. Centrifuge again for 10 min at 45 ×*g*, discard the supernatant, and resuspend each pellet in 5 mL of 0.5 M mannitol. Combine both samples.

4. Take a 100 μL aliquot with a cut pipette tip and determine the protoplast number using a counting chamber. Meanwhile centrifuge the combined protoplasts again for 10 min at 45×*g*. Discard the supernatant and resuspend the pellet in 3 M medium adjusting to a density of 1.2×10^6 protoplasts/mL.
5. Transfer the DNA solution into a glass tube and carefully add 250 μL of the protoplast solution using a cut pipette tip. Add 350 μL of the PEG solution and mix gently by rolling the tube. Incubate the mixture for 30 min at room temperature and mix again every 5 min by gentle rolling. Dilute the mixture with 3 M medium every 5 min adding 1, 2, 3, and 4 mL, respectively, and carefully mix the solution after each step by rolling the tube.
6. Centrifuge for 10 min at 45×*g*, discard the supernatant, and resuspend the protoplasts in 3 mL of regeneration medium. Transfer two times 1.5 mL of the protoplast solution into two wells of a 6-well culture plate. Seal the plate with Parafilm and incubate overnight in the dark at 25°C followed by incubation under normal growth conditions for 10 days. During this time regeneration of protoplasts is initiated.
7. Plate 1 mL of the regenerating protoplasts onto a 9 cm petri dish containing solidified standard growth medium. The medium should be covered with a sterile cellophane sheet which facilitates the transfer of the regenerating plants at subsequent stages. Grow the cultures under standard conditions for 3 days.
8. To select stable transformed plants, transfer the cellophane sheet with the cultures onto solidified growth medium containing the appropriate concentration of the selection antibiotic (depending on the used expression vector) for 2 weeks, followed by a 2-week release period on medium without selection antibiotic and a second selection period of 2 weeks. Plants surviving the second round of selection are considered to be stable transformants.

3.4. PCR-Based Analysis of Putative *P. patens* Transgenic Lines

After selection of regenerating plants, genomic DNA of individual lines can be analyzed by PCR with primers flanking the amiRNA sequence present in the expression construct to identify transgenic lines that had integrated the amiRNA construct.

1. Fill 50 μL of 1X DNA extraction buffer into each well of a 96-well PCR plate. Place 1–5 mg plant material (approximately corresponding to one gametophore or two to three protonema filaments) from each putative transformant into one well containing 50 μL of 1X DNA extraction buffer.
2. Cover the plate using self-adhesive aluminum foil and incubate for 15 min at 45°C in a thermocycler.

3. Directly use 10 μL of the extract for PCR analysis or store it at –20°C until needed.
4. For PCR amplification mix
   10 μL of moss extract
   5 μL of 10X PCR buffer with $(NH_4)_2SO_4$
   4 μL of 3 mM spermidine
   6 μL of 25 mM $MgCl_2$
   5 μL of dNTP mix (2 mM each)
   1 μL of 20 pmol/μL forward primer
   1 μL of 20 pmol/μL reverse primer
   0.5 μL of 5 U/μL Taq DNA polymerase (Promega)
   18.5 μL of H_2O

   The first denaturing step of the PCR is carried out for 5 min at 94°C. In total 40 PCR cycles are recommended. The annealing temperature and the extension time are dependent on the primer sequences and the length of the PCR product.

3.5. RNA Isolation

High-quality total RNA is required for the molecular analysis of transgenic lines including the analysis of (i) amiRNA expression, (ii) amiRNA target expression, and (iii) validation of target cleavage. RNA isolation from *P. patens* using TRIzol reagent (Invitrogen) results in high yields of intact RNA. Furthermore, small RNAs are nicely recovered by this method. We follow the manufacturer's instructions with minor modifications (*see* **Note 3**).

1. Homogenize moss material in liquid nitrogen with mortar and pestle and transfer to a 15 mL Falcon tube (keep the plant material frozen until you add TRIzol).
2. Add 1 mL of TRIzol to 50–100 mg frozen tissue and mix by vortexing.
3. Incubate 5 min at room temperature.
4. Centrifuge 15 min at 5,000×*g* and 4°C.
5. Transfer the supernatant into a new Falcon tube (discard cell debris).
6. Add 0.2 mL of chloroform per initial milliliter of TRIzol, vortex for 15 s.
7. Incubate 5 min at room temperature.
8. Centrifuge 30 min at 5,000×*g* and 4°C.
9. Pipette upper aqueous phase into a new Falcon tube and add an equal volume of isopropanol and mix.
10. Incubate 10 min at 4°C.
11. Centrifuge 20 min at 5,000×*g* and 4°C. Discard the supernatant carefully without losing the RNA pellet.

12. Wash the pellet with 1 mL of 75% ethanol per initial milliliter of TRIzol and centrifuge 8 min at 5,000×*g* and 4°C.
13. Remove the supernatant quantitatively and dry the RNA pellet.
14. Dissolve the RNA in 100–200 μL of RNase-free water. You may incubate the sample for 10 min at 65°C to dissolve the RNA.

3.6. Small RNA Blotting

To prove the correct maturation of the amiRNA from the *A. thaliana* miR319a precursor and its accumulation in the transgenic lines, you should perform small RNA gel blot analysis with an antisense probe for the amiRNA.

1. Preparation of polyacrylamide gel mix: 12 mL of Rotiphorese Gel 40 (37.5:1), 4 mL of 10X TBE buffer, 20 g urea (final concentration, 8.3 M), 320 μL of 10% APS, 42.4 μL of TEMED, adjust the volume to 40 mL with H_2O. We prepare 16 cm × 20 cm gels using 1 mm spacers.
2. Allow gel to polymerize for 1 h.
3. Assemble gel apparatus and add the running buffer (0.5X TBE).
4. Clean wells with running buffer making sure there are no leaks.
5. Load 50–70 μg of total RNA per lane, adjust the volume to 10 μL with DEPC-H_2O.
6. Add 10 μL of 2X RNA loading buffer to your RNA sample.
7. Heat RNA at 65°C for 10 min, chill on ice for 1 min.
8. Load samples and run at 60–80 V until the bromophenol blue reaches the bottom of the gel (~16 h).
9. After electrophoresis, stain the gel in 0.5X TBE containing 0.5 μg/mL of ethidium bromide for 30 min. This will allow you to visualize tRNAs and 5S RNA to check for RNA integrity.
10. Rinse gel in 0.5X TBE to remove excess of ethidium bromide.
11. Cut the Hybond N^+ nylon membrane to a size slightly larger than the gel.
12. Soak the membrane in transfer buffer (0.5X TBE) for 1 min.
13. Soak four pieces of Whatman paper (GB002 grade) in 0.5X TBE.
14. Set up transfer in the Trans-Blot Semi-Dry Electrophoretic Transfer Cell as such from bottom (anode) to top (cathode): two Whatman papers (GB002 grade), Hybond

N+ nylon membrane, gel, two Whatman papers (GB002 grade), and cathode plate. Make sure to roll out any bubbles.

15. Transfer the RNA at 400 mA for 1 h. The voltage will start out low but will increase by the end of the transfer.
16. Wash blot in 0.5X TBE to remove any traces of the gel.
17. Place wet membrane on a wet sheet of filter paper and UV-crosslink at optimal setting. Store membrane dry until use.
18. For prehybridization use 50 mL of hybridization buffer. Prehybridization should be carried out for at least 2 h (*see* **Note 4**).
19. Oligonucleotide labeling reaction:

    1 μL of 10 pmol/μL oligonucleotide

    11 μL of H_2O

    Heat for 5 min at 75°C, cool 3 min on ice

    Add 2 μL of Reaction Buffer A (Fermentas)

    5 μL of 10 μCi/μL [^{32}P]-ATP

    1 μL of 10 U/μL T4 polynucleotide kinase (Fermentas)

    Mix well with pipette and incubate for 60 min at 37°C. Stop the labeling reaction by heat inactivation of the enzyme for 10 min at 68°C.
20. Purify the labeled oligonucleotide using the Nucleotide Removal Kit (Qiagen) and elute the oligonucleotide from the column by eluting twice with 150 μL of EB buffer, combine eluates.
21. Before adding the labeled oligonucleotide to the hybridization buffer, heat for 5 min at 75°C. Before hybridization, exchange the prehybridization buffer and hybridize in 20 mL (*see* **Note 5**) of fresh hybridization buffer for 12–18 h.
22. Wash the membrane: washing is crucial and there are no general rules (washing steps depend strongly on T_m of the hybridization probe, expression level of the amiRNA, etc.). Washes should be performed at a temperature 10°C below the calculated T_m of the oligonucleotide probe.

    Discard hybridization solution. Wash briefly with 2X SSC, 0.2% SDS. Wash 10 min with 2X SSC, 0.2% SDS (check signals on the membrane with a Geiger tube; if necessary continue washing). Wash 5–10 min with 1X SSC, 0.2% SDS (check signals on the membrane with a Geiger tube, if necessary continue washing and increase washing temperature moderately).

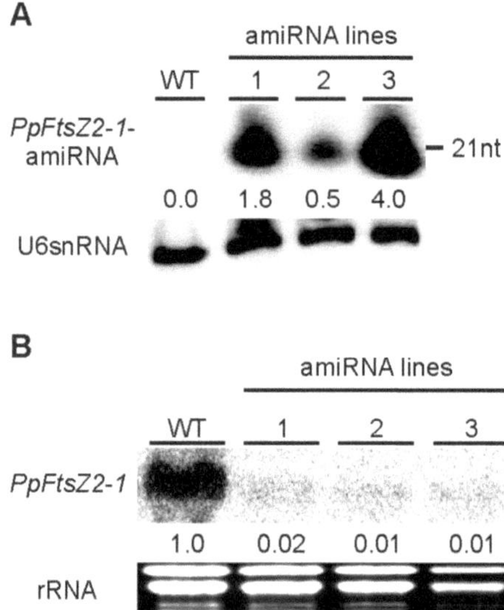

Fig. 5.2. Analysis of *P. patens* lines expressing *PpFtsZ2-1*-amiRNA. (**a**) Expression analysis of *PpFtsZ2-1*-amiRNA expression levels in *P. patens* lines (1–3) harboring the *PpFtsZ2-1*-amiRNA expression construct and in *P. patens* wild type (control). A total of 50 μg of RNA from each line was blotted and hybridized with a *PpFtsZ2-1*-amiRNA antisense probe. Hybridization with an antisense probe for U6 snRNA served as control. *PpFtsZ2-1*-amiRNA expression levels were normalized to the U6 snRNA control. *Numbers* indicate relative *PpFtsZ2-1*-amiRNA expression levels. (**b**) RNA gel blots (20 μg each) from *P. patens* WT and *PpFtsZ2-1*-amiRNA overexpression lines (1–3) hybridized with a *PpFtsZ2-1* probe. The ethidium bromide stained gel below indicates equal loading. The hybridization signals were normalized to the rRNA bands and the *PpFtsZ2-1* expression level in WT was set to one. *Numbers* indicate the relative *PpFtsZ2-1* mRNA levels (modified from Khraiwesh et al. (33); www.plantphysiol.org; Copyright American Society of Plant Biologists).

23. Wrap membrane with Saran wrap and expose to a PhosphorImager to detect hybridization signals (**Fig. 5.2**).
24. If membranes should be used for additional hybridizations (e.g., control hybridization for normalization), the membranes have to be stripped to remove bound radioactively labeled probes. Prepare 0.1% SDS and heat the solution to 95°C. Wash the membrane two times for 15 min at 90°C. Expose the membrane for at least 24 h to control complete removal of the probe. Stripped membranes can be frozen at –20°C wrapped with Saran wrap and can be re-used several times.
25. In order to compare amiRNA expression levels in different transgenic lines, it is strongly recommended to perform an additional hybridization using an oligonucleotide detecting a constitutively expressed RNA. We routinely use an antisense oligonucleotide detecting U6 snRNA that serves as internal standard for sRNA analysis. Hybridization with

this oligonucleotide (5′-GGGGCCATGCTAATCTTCT CTG-3′) allows to control equal loading of the RNA samples and subsequent normalization of amiRNA expression levels (**Fig. 5.2**). Labeling and hybridization are carried out as described above. Quantification of amiRNA and U6 snRNA hybridization signals can be performed using common quantification software (e.g., Quantity One; Bio-Rad).

3.7. Expression Analysis of amiRNA Target Genes by RNA Gel Blots

The expression of an amiRNA should cause amiRNA-mediated cleavage of the intended target transcript and reduced steady-state target RNA levels. The effect of this posttranscriptional silencing and the downregulation of the target RNA have to be confirmed by RNA gel blots or quantitative real-time PCR (qRT-PCR, *see* **Section 3.8**).

1. Clean the gel chamber and combs properly to remove RNases. Leave the gel chamber filled with a detergent solution for 1 h and rinse thoroughly with water afterward.
2. Prepare 150 mL of a 1% denaturing gel for a Midi-Gel (*see* **Note 6**) in the fume hood (formaldehyde is toxic and has to be added in the fume hood): Mix 1.5 g agarose with 108 mL of distilled H_2O, boil until agarose is completely dissolved, cool the solution down, then add 15 mL of 10X MOPS buffer and 27 mL of 37% (w/v) formaldehyde, mix and pour the gel solution into gel tray, and insert comb.
3. Mix RNA sample (up to 30 μg) with 1 volume of RNA denaturation buffer, incubate 10 min at 65°C, cool sample immediately on ice, add 0.1 volume of 10X RNA loading buffer, mix and immediately load onto the gel; until loading keep the samples on ice. If you include RNA markers, these have to be treated in exactly the same way.
4. Electrophoresis is carried out in 1X MOPS buffer (under fume hood).
5. To achieve a good separation of the RNA, run your gel at a constant voltage of 100 V.
6. Let the gel run until the bromophenol blue reaches 2/3 of the gel length.
7. Before blotting take a photo of the gel for documentation (e.g., confirmation of equal loading of the gel).
8. Cut the Hybond N^+ nylon membrane to a size slightly larger than the gel and soak the membrane in distilled water for 15 min.
9. Set up RNA transfer in a Turboblotter system as such from bottom to top: 20 Whatman papers (dry GB004 grade), 4 Whatman papers (dry GB002 grade), 1 Whatman paper

(wet GB002 grade), wet Hybond N+ nylon membrane, RNA gel, 3 Whatman papers (wet GB002 grade), and cover plate. Make sure to roll out any bubbles.

10. Transfer at room temperature for 4 h (additional transfer time may be required for thicker gels).
11. Place wet membrane on a wet sheet of filter paper and UV-crosslink at optimal setting. Store membrane dry until use.
12. To prepare 100 mL of hybridization buffer, mix 34.2 mL of 1 M Na_2HPO_4, 15.8 mL of 1 M NaH_2PO_4, 200 μL of 0.5 M EDTA pH 8.0, 13.8 mL of H_2O, and 35 mL of 20% SDS and heat the solution to 67°C.
13. Boil 1 mL of 10 mg/mL salmon sperm DNA for 10 min, cool on ice, and add to the pre-warmed hybridization buffer.
14. For prehybridization use 50 mL of the hybridization buffer. Prehybridization should be carried out for at least 2 h at 67°C.
15. The labeling of the hybridization probe (cDNA or genomic fragments of the target to be analyzed) is performed using the Rediprime II Random Prime Labeling System (Amersham Biosciences) according to the manufacturer's instructions.
16. Before hybridization, exchange buffer. The volume of buffer has to be kept as small as possible (*see* **Note 5**). Add the denatured radioactively labeled probe to the buffer and hybridize for 12–18 h at 67°C.
17. Prepare the wash solutions and pre-heat solutions to 67°C.
18. Add wash solution (1X SSC, 0.1% SDS) and rinse the membrane briefly, discard into radioactive waste. Wash with 1X SSC, 0.1% SDS for 10 min at 67°C. Wash three times with 0.5X SSC, 0.1% SDS for 10 min at 67°C.
19. Wrap the membrane with Saran wrap and expose to a PhosphorImager to detect hybridization signals (**Fig. 5.2**).
20. If membranes should be used for additional hybridizations (e.g., control hybridization for normalization), they have to be stripped to remove bound radioactively labeled probes. Prepare 0.1% SDS and heat the solution to 95°C. Wash the membrane two times for 15 min at 90°C. Expose the membrane for at least 24 h to control complete removal of the probe. Stripped membranes can be frozen at –20°C wrapped with Saran wrap and can be re-used several times. For control hybridizations to monitor equal loading of RNA samples and subsequent

normalization of RNA expression levels, we routinely use the constitutively expressed gene *PpEF1* encoding a translation elongation factor. Amplify a *PpEF1* cDNA fragment from *P. patens* using the primers 5′-agcgtggtatcacaattgac-3′ and 5′-gatcgctcgatcatgttatc-3′. Use this fragment for radioactive labeling and hybridization of RNA blots as described earlier.

3.8. Expression Analysis of amiRNA Target Genes by Quantitative Real-Time PCR

1. Treat 1 μg of total RNA with 1 U DNase I (Fermentas) and reverse-transcribe into first-strand cDNA using TaqMan Reverse Transcription Reagents (Applied Biosystems, USA) according to the manufacturer's instructions. Include a minus RT control for each RNA probe within your analysis to check for DNA contamination.
2. Design the gene-specific primers according to the following rules (primer design may be performed using special software, e.g., Primer Express, Applied Biosystems): Avoid sequences with consecutive identical nucleotides (i.e., GGGG or CCCC), the amplified product should have a length from ~50 bp to max ~800 bp (the shorter the better), $T_m \geq 60°C$, no more than 2 G or C in the last five nucleotides at the 3′-end. The PCR primers should flank the amiRNA recognition motif. The 5′ and 3′ cleavage products that are generated by miRNA-directed RNA cleavage may be stable and may be amplified by primers that do not flank the miRNA recognition motif. If primers are designed to span the miRNA binding site only intact non-cleaved mRNAs are detected.
3. Design your experiment to include H_2O control, three independent biological replicates, and three technical replicates for each sample.
4. Use *PpEF1α* as constitutively expressed genes for normalization.
5. Adjust the final cDNA concentration to 10 ng/μL and use 50 ng for each PCR reaction. Prepare 150 ng (15 μL) for each sample in a 1.5 mL Eppendorf tube which resembles three technical replicates.
6. For each sample prepare 60 μL of Light Cycler 480 SYBR Green I Master Mix (Roche) according to the manufacturer's instructions and add to 15 μL of cDNA mentioned in Step 5. Mix and spin down briefly.
7. Divide the 75 μL of CR mixture equally into three wells (23 μL/well) of a 96-well real-time PCR plate. Seal the plate, centrifuge briefly, and run PCR in real-time PCR cycler (e.g., Roche 480 Light Cycler).

3.9. Detection of RNA Cleavage Products by RNA Ligase-Mediated Rapid Amplification of cDNA Ends (RLM-RACE)

Besides the analysis of amiRNA target expression the amiRNA-directed cleavage of the target RNA should be confirmed by RLM 5′ RACE. To detect amiRNA-mediated cleavage products we apply a modified protocol of the GeneRacer Kit (Invitrogen). This protocol was originally developed to amplify 5′-ends from full-length RNAs (38). In the original protocol the mRNA is first dephosphorylated to remove 5′ phosphate residues which are present at the 5′-end of RNA degradation products. In a subsequent step, the cap is removed and replaced with an RNA adaptor oligonucleotide (GeneRacer RNA oligonucleotide). During reverse transcription, this RNA adaptor sequence is incorporated into the cDNA. 5′ RACE PCR is then performed using the homologous GeneRacer 5′ primer that is specific to the RNA oligonucleotide sequence and a gene-specific primer. The result is amplified DNA that contains the full-length 5′ cDNA sequence. To detect RNA cleavage products that result from amiRNA-directed RNA cleavage, the protocol is modified (39). The 3′ RNA cleavage products contain phosphorylated 5′-ends that are directly accessible for the RNA adaptor ligation. Furthermore, the detection of full-length capped mRNAs is not intended. Thus, in the modified protocol dephosphorylation and decapping of the RNA are abolished and the RNA adaptor is directly ligated to the RNA samples. During cDNA synthesis the RNA adaptor sequence is incorporated to the cDNA, and subsequent 5′ RACE PCR with an adaptor primer together with a gene-specific primer allows the amplification of the cleavage products.

1. Isolate poly(A)-mRNA from total RNA using Oligotex mRNA Mini Kit (Qiagen) as recommended by the manufacturer.
2. Combine ~150 ng of mRNA and 250 ng of GeneRacer RNA oligonucleotide and incubate at 65°C for 5 min. Set up the ligation reaction by adding

   1 μL of 10 mM ATP

   2 μL of 0.1 mg/mL BSA

   2 μL of 10X RNA ligase buffer

   1 μL of 40 U/μL RNaseOut

   1 μL of 5 U/μL T4 RNA ligase

   Add water up to 20 μL and incubate at 37°C for 1 h.
3. Increase the volume to 100 μL by adding RNase-free water and purify the ligated RNA by phenol:chloroform:isoamyl alcohol (25:24:1) extraction. For this purpose add an equal volume of phenol:chloroform:isoamyl alcohol to the RNA, vortex for 30 s, and centrifuge at 16,000×*g* for 5 min. Transfer the uppermost aqueous phase containing the RNA

into a fresh Eppendorf cap. Precipitate the RNA by adding 0.1 volume of 3 M sodium acetate, pH 5.2, 2.5 volume of ethanol, and GlycoBlue (Applied Biosystems) to reach a final concentration 100 μg/mL. GlycoBlue acts as a carrier during RNA precipitation and facilitates visualization of the pellet. Wash the RNA pellet with 70% ethanol and dissolve in 7 μL of RNase-free water.

4. Start cDNA synthesis using the complete RNA obtained from Step 3 with SuperScript III reverse transcriptase (Invitrogen). Set up the reaction as follows:

   7 μL of adaptor ligated RNA

   1 μL dNTP mix (10 mM each)

   1 μL of 50 pmol/μL Oligo$(dT)_{18}$ primer

   Incubate at 65°C for 5 min, place on ice for 2 min, and continue by adding

   4 μL of 5X RT buffer

   4 μL of 25 mM $MgCl_2$

   2 μL of 0.1 M DTT

   1 μL of 40 U/μL RNaseOut

   Incubate at 42°C for 2 min, then add

   0.5 μL of 200 U/μL SuperScript III RT

   Incubate at 42°C for 50 min. Stop the reaction at 70°C for 15 min, chill on ice, and spin down briefly. Add 1 μL of 2 U/μL RNase H and incubate for 20 min at 37°C.

5. Start 5′ RACE PCR with the GeneRacer primer and a gene-specific primer (*see* **Note 7**). Set up the PCR reaction as follows:

   1 μL of 10 μM GeneRacer 5′ primer

   1 μL of 10 μM reverse gene-specific primer

   1 μL of 5′ RACE cDNA

   3 μL of 10X Pfx Buffer

   0.5 μL dNTP mix (10 mM each)

   0.25 μL of 2.5 U/μL Platinum Pfx DNA Polymerase

   0.5 μL of 50 mM $MgSO_4$

   Add H_2O to 30 μL

   Perform the PCR using the following conditions:

   Step 1: 94°C, 2 min

   Step 2: 94°C, 30 s

   Step 3: 72°C, 1 min per 1 kb (repeat Steps 2 and 3 five times)

   Step 4: 94°C, 30 s

Step 5: 70°C, 1 min per 1 kb (repeat Steps 4 and 5 five times)

Step 6: 94°C, 30 s

Step 7: 60–68°C, 30 s

Step 8: 68–72°C, 1 min per 1 kb (Steps 6–8 are repeated 20 to 25 times)

Step 9: 68–72°C, 10 min

6. Analyze the PCR products by DNA gel electrophoresis. If you observe DNA smear or multiple bands from the 5′ RACE PCR a subsequent nested PCR is recommended. Amplify nested RACE PCR products using a nested gene-specific primer and the GeneRacer 5′ nested primer (*see* **Note 8**). Set up the nested PCR as follows:
   2.5 μL of 10X Pfx Buffer

   0.5 μL dNTPs (10 mM each)

   0.5 μL of 50 mM $MgSO_4$

   1 μL of 10 μM nested GeneRacer 5′ primer

   1 μL of 10 μM reverse nested gene-specific primer

   1 μL from the initial 5′ RACE PCR (template)

   0.25 μL of 2.5 U/μL Platinum Pfx DNA Polymerase

   18.25 μL of H_2O

   PCR conditions:

   Step 1: 94°C, 2 min

   Step 2: 94°C, 30 s

   Step 3: 65°C, 30 s

   Step 4: 68°C, 1 min per kb (Steps 2–4 are repeated 15 to 25 times)

   Step 5: 68°C, 10 min
7. Analyze the PCR products by DNA gel electrophoresis. Elute fragments that correspond to the expected size of the cleavage product, clone, and sequence (*see* **Note 9**).

4. Notes

1. Design amiRNAs lacking perfect sequence complementarity at the 3′-end, as this reduces transitivity.
2. In most of the cases the amiRNA was expressed from endogenous miRNA precursors. The *A. thaliana* miR319a precursor can be used routinely for the expression of amiRNAs in *P. patens*.

3. Be careful working with RNA, as it is easily degraded by RNases which can be carried over to your RNA samples. Use autoclaved pipette tips and baked glassware. Wear gloves when you work with RNA. Prepare all required solutions with DEPC-treated H_2O.
4. The prehybridization temperature and hybridization temperature depend on the sequence of the amiRNA antisense oligonucleotide. Calculate the T_m (annealing temperature) of the oligonucleotide using the following formula (this formula is valid for small oligonucleotides with a size of 17–24 nt): $T_m = (A+T) \times 2 + (C+G) \times 4$. The hybridization temperature should be 10°C below the calculated T_m of the oligonucleotide.
5. Hybridization conditions may differ substantially based on the intention of the designed experiment. The most important criteria are the stringency conditions. High hybridization and wash temperatures coupled with low salt concentrations are high-stringency hybridization conditions. Low temperature and high-salt concentrations favor unspecific hybridizations and represent low-stringency hybridization conditions. Keep the volume of hybridization buffer as small as possible to achieve high probe concentration during the hybridization. The membrane has to be covered completely.
6. To achieve a good separation of RNA samples, we recommend using a Midi-Gel rather than using Mini-Gels. Furthermore, if high amounts of RNA (>10 μg of total RNA) are loaded onto the gel it is better to use a 10- or 12-well comb. When using a 20-well comb, the amount of RNA per well might be too much and may result in bad signals after the hybridization.
7. Gene-specific primers are designed with 50–70% GC content and having high annealing temperature (>72°C).
8. Nested gene-specific primers, located upstream of the initial gene-specific RACE PCR primer, should have a 50–70% GC content and an annealing temperature above 68°C.
9. It is recommended to sequence several independent clones to map the amiRNA cleavage site.

Acknowledgments

We gratefully acknowledge financial support from the German Academic Exchange Service (DAAD; Ph.D. fellowships to I.F. and M.A.A.).

References

1. de Carvalho, F., Gheysen, G., Kushnir, S., Van Montagu, M., Inze, D., and Castresana, C. (1992) Suppression of β-1,3-glucanase transgene expression in homozygous plants. *EMBO J.* **11**, 2595–2602.
2. Lee, R. C., Feinbaum, R. L., and Ambros, V. (1993) The *C. elegans* heterochronic gene *lin-4* encodes small RNAs with antisense complementarity to *lin-14*. *Cell* **75**, 843–854.
3. Matzke, M. A., Primig, M., Trnovsky, J., and Matzke, A. J. (1989) Reversible methylation and inactivation of marker genes in sequentially transformed tobacco plants. *EMBO J.* **8**, 643–649.
4. Napoli, C., Lemieux, C., and Jorgensen, R. (1990) Introduction of a chimeric chalcone synthase gene into *Petunia* results in reversible co-suppression of homologous genes *in trans*. *Plant Cell* **2**, 279–289.
5. Tomari, Y. and Zamore, P. D. (2005) Perspective: machines for RNAi. *Genes Dev.* **19**, 517–529.
6. Hamilton, A. J. and Baulcombe, D. C. (1999) A species of small antisense RNA in posttranscriptional gene silencing in plants. *Science* **286**, 950–952.
7. Bartel, D. P. (2004) MicroRNAs: genomics, biogenesis, mechanism, and function. *Cell* **116**, 281–297.
8. Chapman, E. J. and Carrington, J. C. (2007) Specialization and evolution of endogenous small RNA pathways. *Nat. Rev. Genet.* **8**, 884–896.
9. Kurihara, Y. and Watanabe, Y. (2004) Arabidopsis micro-RNA biogenesis through Dicer-like 1 protein functions. *Proc. Natl. Acad. Sci. USA* **101**, 12753–12758.
10. Jones-Rhoades, M. W., Bartel, D. P., and Bartel, B. (2006) MicroRNAs and their regulatory roles in plants. *Annu. Rev. Plant. Biol.* **57**, 19–53.
11. Axtell, M. J., Snyder, J. A., and Bartel, D. P. (2007) Common functions for diverse small RNAs of land plants. *Plant Cell* **19**, 1750–1769.
12. Fattash, I., Voss, B., Reski, R., Hess, W. R., and Frank, W. (2007) Evidence for the rapid expansion of microRNA-mediated regulation in early land plant evolution. *BMC Plant Biol.* **7**, 13.
13. Aravin, A. A., Lagos-Quintana, M., Yalcin, A., Zavolan, M., Marks, D., Snyder, B., Gaasterland, T., Meyer, J., and Tuschl, T. (2003) The small RNA profile during *Drosophila melanogaster* development. *Dev. Cell* **5**, 337–350.
14. Ossowski, S., Schwab, R., and Weigel, D. (2008) Gene silencing in plants using artificial microRNAs and other small RNAs. *Plant J.* **53**, 674–690.
15. Allen, E., Xie, Z., Gustafson, A. M., and Carrington, J. C. (2005) microRNA-directed phasing during *trans*-acting siRNA biogenesis in plants. *Cell* **121**, 207–221.
16. Peragine, A., Yoshikawa, M., Wu, G., Albrecht, H. L., and Poethig, R. S. (2004) *SGS3* and *SGS2/SDE1/RDR6* are required for juvenile development and the production of *trans*-acting siRNAs in *Arabidopsis*. *Genes Dev.* **18**, 2368–2379.
17. Vazquez, F., Vaucheret, H., Rajagopalan, R., Lepers, C., Gasciolli, V., Mallory, A. C., Hilbert, J. L., Bartel, D. P., and Crete, P. (2004) Endogenous *trans*-acting siRNAs regulate the accumulation of *Arabidopsis* mRNAs. *Mol. Cell* **16**, 69–79.
18. Yoshikawa, M., Peragine, A., Park, M. Y., and Poethig, R. S. (2005) A pathway for the biogenesis of *trans*-acting siRNAs in *Arabidopsis*. *Genes Dev.* **19**, 2164–2175.
19. Axtell, M. J., Jan, C., Rajagopalan, R., and Bartel, D. P. (2006) A two-hit trigger for siRNA biogenesis in plants. *Cell* **127**, 565–577.
20. Fahlgren, N., Montgomery, T. A., Howell, M. D., Allen, E., Dvorak, S. K., Alexander, A. L., and Carrington, J. C. (2006) Regulation of *AUXIN RESPONSE FACTOR3* by *TAS3* ta-siRNA affects developmental timing and patterning in *Arabidopsis*. *Curr. Biol.* **16**, 939–944.
21. Hunter, C., Willmann, M. R., Wu, G., Yoshikawa, M., de la Luz Gutierrez-Nava, M., and Poethig, S. R. (2006) Trans-acting siRNA-mediated repression of ETTIN and ARF4 regulates heteroblasty in *Arabidopsis*. *Development* **133**, 2973–2981.
22. Guo, H. S., Xie, Q., Fei, J. F., and Chua, N. H. (2005) MicroRNA directs mRNA cleavage of the transcription factor *NAC1* to downregulate auxin signals for *Arabidopsis* lateral root development. *Plant Cell* **17**, 1376–1386.
23. Vaucheret, H., Vazquez, F., Crete, P., and Bartel, D. P. (2004) The action of *ARGONAUTE1* in the miRNA pathway and its regulation by the miRNA pathway are crucial for plant development. *Genes Dev.* **18**, 1187–1197.
24. Alvarez, J. P., Pekker, I., Goldshmidt, A., Blum, E., Amsellem, Z., and Eshed, Y. (2006) Endogenous and synthetic microRNAs stimulate simultaneous, efficient, and

localized regulation of multiple targets in diverse species. *Plant Cell* **18**, 1134–1151.

25. Niu, Q. W., Lin, S. S., Reyes, J. L., Chen, K. C., Wu, H. W., Yeh, S. D., and Chua, N. H. (2006) Expression of artificial microRNAs in transgenic *Arabidopsis thaliana* confers virus resistance. *Nat. Biotechnol.* **24**, 1420–1428.
26. Parizotto, E. A., Dunoyer, P., Rahm, N., Himber, C., and Voinnet, O. (2004) In vivo investigation of the transcription, processing, endonucleolytic activity, and functional relevance of the spatial distribution of a plant miRNA. *Genes Dev.* **18**, 2237–2242.
27. Schwab, R., Ossowski, S., Riester, M., Warthmann, N., and Weigel, D. (2006) Highly specific gene silencing by artificial microRNAs in *Arabidopsis. Plant Cell* **18**, 1121–1133.
28. Warthmann, N., Chen, H., Ossowski, S., Weigel, D., and Herve, P. (2008) Highly specific gene silencing by artificial miRNAs in rice. *PLoS ONE* **3**, e1829.
29. Zeng, Y., Wagner, E. J., and Cullen, B. R. (2002) Both natural and designed micro RNAs can inhibit the expression of cognate mRNAs when expressed in human cells. *Mol. Cell* **9**, 1327–1333.
30. Boden, D., Pusch, O., Silbermann, R., Lee, F., Tucker, L., and Ramratnam, B. (2004) Enhanced gene silencing of HIV-1 specific siRNA using microRNA designed hairpins. *Nucleic Acids Res.* **32**, 1154–1158.
31. Dickins, R. A., Hemann, M. T., Zilfou, J. T., Simpson, D. R., Ibarra, I., Hannon, G. J., and Lowe, S. W. (2005) Probing tumor phenotypes using stable and regulated synthetic microRNA precursors. *Nat. Genet.* **37**, 1289–1295.
32. Qu, J., Ye, J., and Fang, R. (2007) Artificial microRNA-mediated virus resistance in plants. *J. Virol.* **81**, 6690–6699.
33. Khraiwesh, B., Ossowski, S., Weigel, D., Reski, R., and Frank, W. (2008) Specific gene silencing by artificial microRNAs in *Physcomitrella patens*: An alternative to targeted gene knockouts. *Plant Physiol.* **148**, 684–693.
34. Schwab, R., Palatnik, J. F., Riester, M., Schommer, C., Schmid, M., and Weigel, D. (2005) Specific effects of microRNAs on the plant transcriptome. *Dev. Cell* **8**, 517–527.
35. Mallory, A. C., Reinhart, B. J., Jones-Rhoades, M. W., Tang, G., Zamore, P. D., Barton, M. K., and Bartel, D. P. (2004) MicroRNA control of *PHABULOSA* in leaf development: importance of pairing to the microRNA 5′ region. *EMBO J.* **23**, 3356–3364.
36. Koncz, C., Martini, N., Mayerhofer, R., Koncz-Kalman, Z., Korber, H., Redei, G. P., and Schell, J. (1989) High-frequency T-DNA-mediated gene tagging in plants. *Proc. Natl. Acad. Sci. USA* **86**, 8467–8471.
37. Frank, W., Decker, E. L., and Reski, R. (2005) Molecular tools to study *Physcomitrella patens. Plant Biol.* **7**, 220–227.
38. Volloch, V., Schweitzer, B., and Rits, S. (1994) Ligation-mediated amplification of RNA from murine erythroid cells reveals a novel class of β globin mRNA with an extended 5′-untranslated region. *Nucleic Acids Res.* **22**, 2507–2511.
39. Llave, C., Xie, Z., Kasschau, K. D., and Carrington, J. C. (2002) Cleavage of *Scarecrow-like* mRNA targets directed by a class of *Arabidopsis* miRNA. *Science* **297**, 2053–2056.

Chapter 6

Virus-Induced Gene Silencing in Ornamental Plants

Cai-Zhong Jiang, Jen-Chih Chen, and Michael Reid

Abstract

Virus-induced gene silencing (VIGS) provides an attractive tool for high-throughput analysis of the functional effects of gene knockdown. Virus genomes are engineered to include fragments of target host genes, and the infected plant recognizes and silences the target genes as part of its viral defense mechanism. The consequences of gene inactivation, even of key metabolic, regulatory, or embryo-lethal genes, can thus be readily analyzed. A number of viral vectors have been developed for VIGS; one of the most frequently employed is based on tobacco rattle virus (TRV) due to its wide host range, efficiency, ease of application, and limited disease symptoms. TRV-based VIGS comprises two vectors. One (RNA2) includes a multiple cloning site into which fragments of target genes can be inserted. We have shown that the TRV/VIGS system can simultaneously silence as many as five independent genes. TRV is a mosaic-type virus, and silencing also occurs in a mosaic pattern. It is therefore desirable to have a reporter that can show where target genes have been silenced. The photobleaching induced by silencing phytoene desaturase (PDS) and the loss of purple pigmentation induced by silencing chalcone synthase (CHS) have successfully been used to indicate the location of coordinate silencing of other target genes. In this chapter, we outline our protocols for the use of VIGS for analysis of gene function, focusing particularly on the use of TRV with petunia and tomato.

Key words: Abscission, chalcone synthase, petunia, phytoene desaturase, senescence, tobacco rattle virus, tomato.

1. Introduction

Post-transcriptional gene silencing (PTGS) is one of the most adaptable and specific mechanisms for protection of the genome and elimination of foreign DNA or RNA and culminates in the sequence-specific degradation of so-called aberrant RNA. This highly conserved process was first identified in plants by the apparently bizarre silencing of chalcone synthase (CHS) in petunia

H. Kodama, A. Komamine (eds.), *RNAi and Plant Gene Function Analysis*, Methods in Molecular Biology 744,
DOI 10.1007/978-1-61779-123-9_6, © Springer Science+Business Media, LLC 2011

plants that had been engineered to overexpress this gene (1). PTGS has now been shown to function as an endogenous defense mechanism against viruses by directly targeting the replicative form of the virus (2). During replication of the virus, chimeric double-stranded intermediates are produced. The plant cell recognizes these intermediates as foreign invaders, and a specialized enzyme (Dicer) degrades the double-stranded RNA into small interfering oligonucleotides (siRNA). The single strands of siRNA molecules, bound to an RNA-induced silencing complex (RISC), serve as specific templates that target any transcripts with identical or highly similar sequences for degradation. In virus-induced gene silencing (VIGS), this mechanism is co-opted to target host mRNAs by the simple expedient of including fragments of target host genes into a modified viral genome. The viral silencing mechanism does not differentiate these fragments from viral sequences, and this provides a simple means to downregulate host gene expression (3). The technique has been extended to model plants such as *Arabidopsis thaliana* (4) and crops such as tomato (*Solanum lycopersicum*) (5, 6), potato (*Solanum tuberosum*) (7), barley (8, 9), wheat (10–12), rice and maize (13), *Jatropha* (14), and soybean (15–17). Our laboratory has exploited VIGS for testing gene function in ornamental plants including petunia, tomato, *Impatiens*, chrysanthemum, and *Mirabilis jalapa* (18–20).

There are several advantages to VIGS over gene silencing methods involving transgenic plants expressing inverted repeat constructs (RNAi approaches). First of all, the constructs can easily be generated by directly cloning gene fragments into the viral vector without the need for generating inverted repeats. Second, the VIGS phenotype can be observed in a relatively short time (as short as 10 days in petunia and tomato) after inoculation. VIGS offers an attractively quick method for loss-of-function assay and avoids the need for the time-consuming (and sometimes problematic) processes of transformation and regeneration. Third, VIGS can be used to downregulate proteins encoded by multigene families by using a fragment from a highly conserved region common to all members of the family or by including multiple fragments from the different genes in a single vector. Last, VIGS can reveal the mature phenotypes of embryo-lethal disruptions of gene function.

Many plant viruses have been used to develop VIGS vectors. Early vectors included tobacco mosaic virus (TMV) (3, 9), potato virus X (PVX) (21–24), and tomato golden mosaic virus (TGMV) (25). Some of the viruses may cause strong infection symptoms including chlorosis, leaf distortion, and necrosis (26). A few of them are incapable of infecting the apical meristem and are therefore unlikely to provide information about genes involved in the identity and development of plant tissues and organs (26). The tobacco rattle virus vectors described by Ratcliff et al. (26) have proved to be very useful for VIGS in a wide range of plant taxa.

More recently, a range of other VIGS vectors have been developed, including the *Tomato yellow leaf curl China virus* (27), *African cassava mosaic virus* for VIGS in cassava (28), *Cotton leaf crumple virus* for cotton (29), *Apple latent spherical virus* (16), *Pea early browning virus* (30), and *Bean pod mottle virus* for legumes (31). For monocotyledonous plants, VIGS vectors have been constructed from *Barley stripe mosaic virus* (BSMV) and *Brome mosaic virus* (BMV). These have been widely used in barley, wheat, rice, and maize for gene functional analysis (8, 10–13).

Tobacco rattle virus (TRV)-based vectors are useful for VIGS in a range of plant species due to the wide host range of TRV (5, 26, 32) and overcome many of the difficulties associated with PVX, TMV, and TGMV. For example, the TRV vector induces only very mild symptoms, infects large areas of adjacent cells, and silences expression of genes in vegetative and floral meristems. The TRV-based VIGS system has been shown to function effectively in *Arabidopsis*, *Nicotiana benthamiana*, tomato (5, 6, 32), chrysanthemum, *Impatiens*, *Mirabilis jalapa* (Jiang and Reid, unpublished), *Aquilegia* (33), *Jatropha* (14), and California poppy (34).

TRV is a two-particle positive-sense RNA rod-type virus. RNA1 encodes two replicase proteins, a movement protein, and a cysteine-rich protein. RNA2 encodes the coat protein and two non-structural proteins. Because TRV RNA1 can replicate and move systemically in the plant in the absence of RNA2, researchers reasoned that they could modify RNA2 to facilitate the insertion of fragments of genes targeted for silencing. They constructed binary transformation vectors using T-DNA from *Agrobacterium*, containing 35S promoters and the RNA1 and RNA2 of TRV (5, 26, 32). The 35S promoters stimulate initial transcription of the viral genomes, thus ensuring rapid replication once the plant cells have been infected with the transformed *Agrobacterium*. RNA2 was modified by replacing the genes encoding non-structural proteins with a multiple cloning site (MCS) into which fragments of target genes can readily be inserted. In our studies we have used the vector (**Fig. 6.1**) described by Liu et al. (5, 32).

Although vector construction is outside the scope of the present protocol, it is apparent that continued research will reveal vectors appropriate to rapid evaluation of silencing in a wide range of target species. Most researchers will be able to obtain appropriate vectors from colleagues who have reported them.

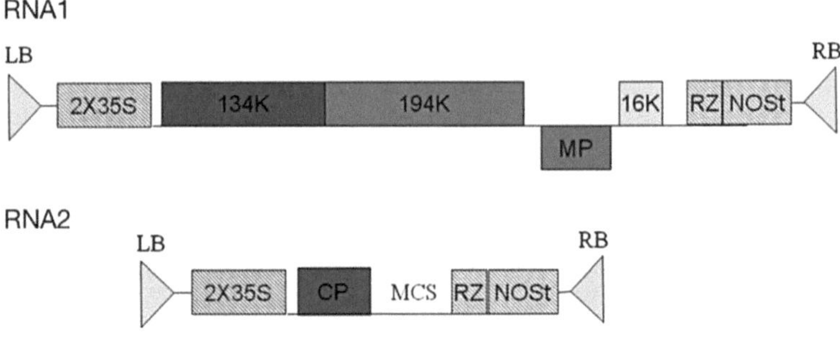

Fig. 6.1. Map for pTRV RNA1 and RNA2. Multiple cloning sites include restriction enzymes of *Eco*RI, *Xba*I, *Stu*I, *Nco*I, *Bam*HI, *Kpn*I, *Sac*I, *Mlu*I, *Xho*I, *Srf*I, and *Sma*I (32).

2. Materials

2.1. Amplification of Target Genes

1. cDNA synthesized from total RNA isolated from relevant plant tissues.
2. Polymerase chain reaction (PCR) primers.
3. Qiagen PCR kit.
4. SuperScript III Reverse Transcriptase (Invitrogen).
5. Agarose.
6. SYBR Safe DNA gel stain (10,000X, Invitrogen).
7. TAE buffer (50X stock): 2 M Tris, 1 M glacial acetic acid, 50 mM Na_2EDTA, pH 8.0.
8. TE buffer: 10 mM Tris–HCl, pH 8.0, 1 mM Na_2EDTA.
9. DNA loading buffer: 15 g ficoll 400, 0.5 g sodium dodecyl sulfate (SDS), 0.25 g bromophenol blue. Bring to 100 mL with TE buffer. Store at 4°C.
10. Gel extraction kit (e.g., QIAEX II Gel Extraction Kit), if required.
11. TA-vector cloning kit (e.g., Promega pGEM-T Easy Vector System).
12. Plasmid DNA miniprep kit (Qiagen).
13. *Escherichia coli* cells (e.g., DH5α). Store at –80°C.
14. SOC broth: 20.0 g Bacto tryptone, 5.0 g Bacto yeast extract, 0.6 g NaCl, 0.5 g KCl. Bring to 1 L with deionized water. Autoclave, cool, then add 10 mL of 1 M $MgCl_2$, 10 mL of 1 M $MgSO_4$, and 20 mL of 20 mM glucose (all three solutions sterilized through a 0.45 μm disposable filter).

15. Luria–Bertani broth (LB): 10.0 g bacteriological tryptone (Difco), 5.0 g yeast extract, 10.0 g NaCl. Bring to 1 L with deionized water and adjust pH to 7 using 1 M NaOH.
16. Ampicillin stock solution (50 mg/mL, 1000X): 0.25 g ampicillin; bring to 5 mL with deionized water, sterile filter. Store at –20°C.
17. Water bath.
18. Incubator preset to 37°C.
19. Thermal cycler (Applied Biosystems).
20. Gel electrophoresis systems.

2.2. Cloning into the Vector

1. *See* **Section 2.1** for agarose, SYBR Safe DNA gel stain, TAE buffer, DNA loading buffer, gel extraction kit, *E. coli* competent cells, SOC broth, LB agar plates with ampicillin, LB broth, ampicillin, and plasmid miniprep kit.
2. Restriction nucleases (New England Biolabs): *Eco*RI, *Xba*I, *Nco*I, *Bam*HI, *Kpn*I, *Sac*I, *Xho*I, *Sma*I. See the MCS of RNA2 (**Fig. 6.1**).
3. pTRV RNA2 vector (5, 32). Vectors are available from Arabidopsis Biological Resource Center at Ohio State University on request and their maps are presented in **Fig. 6.1**.
4. T4 DNA ligase and 10X ligation buffer.
5. DNA clean-up kit (e.g., QIAEX II Gel Extraction Kit that is also adapted for this purpose).
6. Shrimp alkaline phosphatase (SAP) and its corresponding buffer (e.g., Fermentas).
7. 50 mg/mL kanamycin stock solution (1000X).
8. LB agar plates containing 50 μg/mL kanamycin.

2.3. Agrobacterium Transformation

1. *Agrobacterium tumefaciens* strain GV3101 (5, 32).
2. YEP medium: 10 g yeast extract, 10 g Bacto peptone, 5 g NaCl. Adjust pH to 7.0 and bring final volume to 1 L with deionized water. Autoclave.
3. Gentamicin stock solution (20 mg/mL, 1000X): 0.1 g gentamicin; bring to 5 mL with deionized water, sterile filter. Store at –20°C.
4. Kanamycin stock solution.
5. MicroPulser Electroporator (Bio-Rad).

2.4. Inoculation

1. Seedlings of a competent cultivar of the target plant species (*see* **Note 1**). We use petunia and tomato as easily manipulated model systems to study plant senescence and abscission.

2. Luria–Bertani broth (LB): 10 g/L tryptone, 5 g/L yeast extract, 10 g/L NaCl, pH 7.0, with 20 μg/mL gentamicin and 50 μg/mL kanamycin.
3. Inoculation buffer: 10 mM $MgCl_2$, 10 mM MES, 200 μM acetosyringone.
4. 10 mM $MgCl_2$.
5. Centrifuges (Beckman).
6. Shaker (Fisher Scientific).
7. Incubators (28°C).
8. Temperature-controlled growth chamber.
9. 1 mL syringe, needleless (for syringe infiltration, Fisher Scientific).
10. Latex gloves (for syringe infiltration, Fisher Scientific).

3. Methods

3.1. Identification of Target Gene Sequences for Silencing Using VIGS

Sequence fragments (100 bp minimum) of genes of interest (GOIs) should be identical (or at least very highly homologous) to the gene in the plant being used for silencing (*see* **Note 2**). Fragments encoding multiple genes, up to a total of 1.5 kb in length, may be used in the same vector (*see* **Note 3**). Simultaneous silencing of homologous genes requires selection of a silencing fragment from a highly conserved region of the GOIs (*see* **Note 4**).

3.2. Amplification of Target Genes

1. Fragments of the GOI suitable for VIGS should be amplified from cDNA sources using standard PCR amplification procedures (we use the Qiagen PCR kit and follow the included protocol) and gene-specific primers designed to generate a 100–500 bp fragment (*see* **Note 5**).
2. The first-strand cDNA is synthesized using total RNA, oligo dT primer, random hexamer, and SuperScript III Reverse Transcriptase (Invitrogen). It is critical that the fragment be highly homologous with a region of the host-plant gene, and it is therefore important to amplify the fragment from the target species (*see* **Note 6**).
3. Check the PCR products by electrophoresis in 1% agarose gel to ensure that a PCR product of the expected size was specifically amplified.
4. Excise the DNA fragments and gel purify them using the Qiagen DNA purification kit.
5. Clone each PCR product in a TA-vector by following the manufacturer's recommendation (Promega).

6. Transform *E. coli* competent cells with the ligation product. Allow frozen competent *E. coli* cells to thaw slowly on ice. Add 200 μL of the cells to ligation mix and incubate on ice for 30 min. Heat-shock cells for 45 s in a water bath at 42°C and chill them for 5 min on ice.
7. Add 1 mL of SOC broth and incubate for 1 h at 37°C with shaking.
8. Plate 200 μL of cells on LB agar plates containing 50 μg/mL ampicillin. Centrifuge the remaining cells and remove most of the supernatant (leave 100–200 μL in the tube). Resuspend cells in the remaining supernatant and plate them on LB agar plates with ampicillin. Incubate the plates at 37°C overnight.
9. To facilitate subsequent cloning into the pTRV2 vector, the presence of insert DNA is checked by sequencing.

3.3. Cloning into the Vector

Clone the confirmed candidate gene fragments into the multiple cloning site (MCS) of the pTRV RNA2 vector (**Fig. 6.1**).

1. For insertion of target sequence, digest the TA-vector containing the PCR product (**Section 3.2**, Step 9) with *Eco*RI.
2. Open the vector pTRV RNA2 with *Eco*RI, then add 1 μL SAP (*see* **Section 2.2**, Step 6) to the digest to prevent self-ligation.
3. Run the entire digested samples on a 1% agarose gel.
4. Cut out the bands of interest with a clean razor blade and purify the appropriate DNA fragments using Qiagen gel extraction kit.
5. Ligate 30–75 ng of *Eco*RI-digested vector pTRV RNA2 with approximately threefold molar excess of the fragment of GOI with T4 DNA ligase in 1X ligation buffer and a final volume of 20 μL overnight at 15°C.
6. Use the ligated products to transform competent *E. coli* (DH5α, Invitrogen) cells by heat shock (**Section 3.2**, Steps 6–8) and plate them on selective LB media containing antibiotics (50 μg/mL kanamycin).
7. Screen drug-resistant colonies by PCR, using gene-specific primers, to confirm the presence of the correct gene fragments in the constructs (*see* **Note 7**).

3.4. Agrobacterium Transformation

Viral infection is achieved by *Agrobacterium*-mediated infection of seedlings.

1. Transform separate cultures of *A. tumefaciens* strain GV3101 with RNA1 (pTRV RNA1 construct) and the modified RNA2 (pTRV RNA2 construct) by electroporation (18).

2. Select recombinant *Agrobacterium* clones on LB containing 50 μg/mL kanamycin and 20 μg/mL gentamicin. The transformed cells are Gent and Kan resistant.
3. Use PCR to confirm the presence of pTRV2 carrying the candidate gene fragments in the selected transformed cells.
4. Culture the bacteria overnight at 28°C with shaking in YEP medium containing gentamicin (20 μg/mL) and kanamycin (50 μg/mL).
5. Harvest the *Agrobacterium* cells by centrifugation at 3000×g for 15 min and resuspend them in inoculation buffer to an OD_{600} of 2–4.
6. Incubate the cells at room temperature with gentle shaking for a minimum of 3 h to ensure full activation of the *Agrobacterium Vir* genes (*see* **Note 8**).
7. The bacteria containing RNA1 and those with the modified RNA2 are then mixed together in a 1:1 ratio (**Fig. 6.2**) immediately before inoculation.

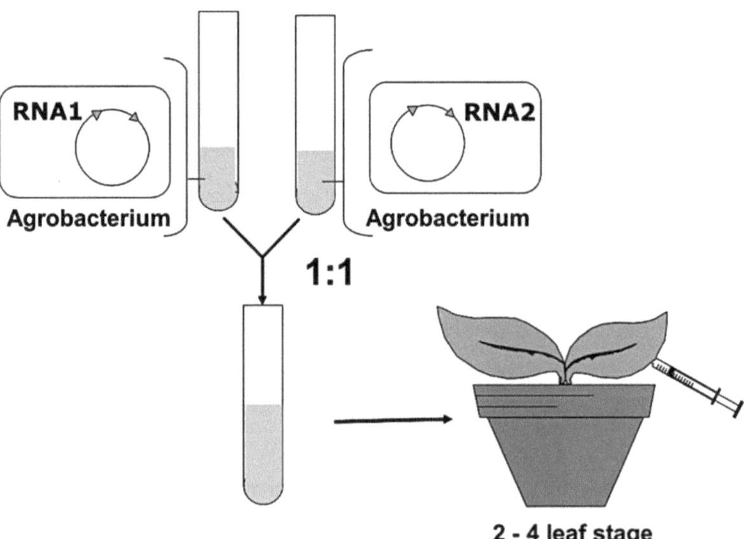

Fig. 6.2. The VIGS experimental system. Separate *Agrobacterium* cultures are transformed with the two viral vectors (RNA1 and RNA2); the RNA2 vector contains a fragment of the gene(s) to be silenced. Following growth of the transformed *Agrobacterium*, the combined cultures are used to infect primary leaves of young petunia plants, by infiltration through the abaxial leaf surface, using a needleless disposable syringe.

3.5. Choice of Plant Cultivar

Test efficacy of silencing, using test gene fragments (PDS, CHS, GFP, etc.), on a range of cultivars of the target species to choose a suitable candidate cultivar (*see* **Note 1**).

3.6. Growing Plants for VIGS

1. Germinate seeds in standard potting medium in a pot. Cover the pots with a clear plastic dome to prevent dehydration and maintain adequate moisture.
2. Grow plants (petunia and tomato) in growth chambers with a 16 h light/8 h dark cycle and a day/night temperature regime of 22/18°C.
3. Plants used for inoculation by leaf infiltration should be 2 weeks old or when the first two true leaves have emerged. Ten-day-old seedlings can be used for vacuum or pressure infiltration.

3.7. Inoculation

We have tested a range of techniques for inoculation; choice of method, which depends on plant species, cultivar, and growth stage, is an important factor in the efficiency of silencing (*see* **Note 9**). For petunia, tobacco, and tomato, we have found leaf inoculation of primary leaves to be very effective.

1. The primary leaves of young plantlets are inoculated by injection of the mixed bacterial culture using a 1 ml disposable syringe without a needle (*see* **Note 10**).
2. Wear latex gloves.
3. Load the *Agrobacterium* mixture (*see* **Section 3.4**, Step 7) into a 1 mL needleless syringe.
4. If needed, gently nick the abaxial surface of the first two true leaves of the seedlings.
5. Place the syringe on the nick side and gently inject the solution into the leaf until the leaves are completely infiltrated.
6. To prevent cross-contamination, replace gloves and syringe for each infiltration when silencing different GOIs.
7. Cover the infiltrated plants with a clear plastic dome or Saran wrap for 2 days to allow quick recovery.
8. Remove the cover and transfer the plants to a temperature-controlled growth chamber.

3.8. Plant Growth Conditions

Temperature conditions during plant growth after inoculation are critical to the efficiency, uniformity, and spread of gene silencing (*see* **Note 11**). For petunia and tomato, we grow the plants under high artificial light (approximately 300 μE/m/s), long photoperiods (16 h day), and a day/night temperature regime of 22/18°C.

3.9. Evaluation of Silencing

Our primary evaluation tool is the visible symptoms of silencing of PDS (photobleaching of the leaves) or of CHS (white floral spots, sectors, or corollas on normally blue or violet flowers) (**Fig. 6.3**). To confirm that silencing has been achieved we

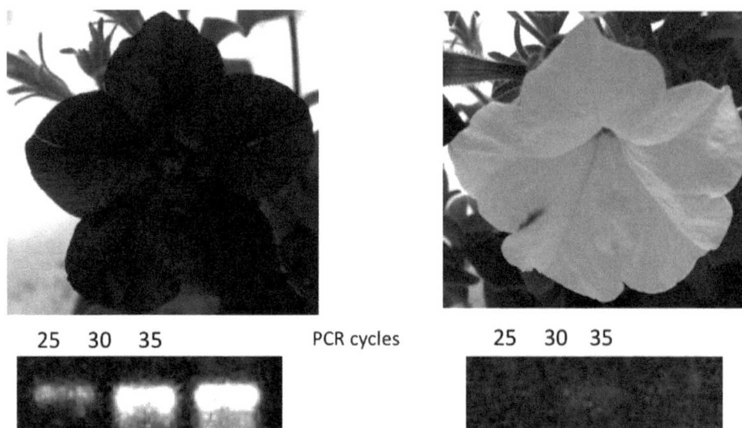

Fig. 6.3. The effect of silencing chalcone synthase (CHS) on flower color and CHS transcript abundance in petunia. Plants were infected with TRV containing a fragment of the petunia CHS gene. Petals were harvested and transcript abundance was visualized by RT-PCR amplification of CHS in total RNA using primers binding to a region outside the silencing fragment.

typically measure transcript abundance using semi-quantitative or real-time RT-PCR (18, 19).

1. Total RNA is extracted using plant RNA isolation TRIzol Reagent (Invitrogen).
2. The RNA is treated with RNase-free DNase (Promega, Madison, WI, USA) to remove any contaminating genomic DNA.
3. First-strand cDNA is synthesized using the SuperScript III kit (Invitrogen).
4. PCR primers for amplifying the transcripts must be designed outside the region targeted for gene silencing to avoid amplification of the fragment included in the viral RNA2 construct.

3.10. Use of Reporter Genes for Evaluation of Silencing Efficiency

Because VIGS typically results in chimeric plants with some portions uninfected, and therefore unsilenced, it is essential to incorporate the silencing of a reporter gene as a tool to indicate sites of silencing (*see* **Note 12**). In our studies of floral senescence, we use a fragment of an endogenous petunia chalcone synthase gene (*CHSJ*; GenBank accession number X14599) in the silencing construct. Silencing (**Fig. 6.3**) is seen as white spots, sectors, and flowers on infected plants (18).

3.11. Future Perspectives

Although VIGS seldom results in 100% suppression of the target gene at transcript level, our experience and the data that we have obtained demonstrate the value of VIGS, based on TRV vectors, as a tool for reverse genetic analysis in petunia and tomato. In our laboratory, infection routinely results in a very high frequency of silencing symptoms. We are using this tool to examine the effects of silencing a wide range of transcription factors on flower senescence, abscission, and other aspects of plant growth and development. Among the technical questions that remain to be answered is the maximum number of independent transcripts that can be silenced using a single vector. With more transcript fragments per vector, the throughput of the VIGS technique increases, opening the possibility of genome-wide functional analysis. Another interesting avenue worth exploring is the use of VIGS to silence all members of a multi-gene family, thereby generating information on the function of genes whose redundancy makes functional analysis challenging using other reverse genetic approaches. Petunia and tomato have proven to be excellent model plants for the application of VIGS, and the technique is increasingly being used in a range of other ornamental, model, and crop species. The wide host range of TRV suggests that the technique will be applicable in many taxa. Some taxa have proved to be recalcitrant, and further research is required to find strategies to allow the application of VIGS in such cases. As an example, we were unable until recently to demonstrate VIGS in *Mirabilis jalapa*. This species contains a potent anti-viral protein; in recent studies (Jiang, Liang, Singh, and Reid, unpublished) we showed that co-silencing of the anti-viral protein dramatically increased silencing efficiency.

4. Notes

1. We initially used the homozygous model petunia cultivar V26 in our studies of silencing floral genes, but under our experimental conditions the silencing phenotype was limited to diffuse small white spots on the corollas and did not lend itself to biochemical or physiological characterization (18). We therefore tested the effects of silencing CHS on a range of purple-flowered commercial cultivars (Goldsmith Seeds, Gilroy, CA, USA) and found significant variations in the silencing phenotype. In some, clear silenced sectors and whole silenced corollas were common; others showed polka dot and/or diffuse silencing patterns similar to that seen in V26. The genetic basis for these differences is unknown – possible explanations include differences in movement of either the virus or the silencing signal between cells. This phenomenon is not restricted to petunia. In studies with silencing PDS in tomato we have also seen cultivar-dependent variation in the silencing

phenotype (19). Whatever the cause, it is obviously important for investigators to choose a cultivar that responds optimally to VIGS.

2. We find that silencing is highly efficient with fragments in the 150 bp range. Lu et al. (23, 24) suggested that homologous inserts as short as 23 nucleotides are sufficient to induce gene silencing, although the silencing is less extensive and more transient than with inserts with longer homologous sequences. This implies that having inserts with short regions of very high identity with the target gene is more important than having a larger fragment with moderate overall similarity.
3. The RNA2 vector can accept inserts up to 1.5 kb, allowing several gene fragments to be placed in the same vector and silenced simultaneously. Thus, infection of purple-flowered cultivars with TRV containing fragments of both PDS and CHS resulted in leaf photobleaching and white patterns on the flowers. The abundance of both transcripts was reduced, confirming simultaneous silencing of both genes at the same location (18). Using fragments from multiple members of a petunia multigene family, we were able to show that as many as five genes (including CHS) could be silenced simultaneously (Hunter, Jiang, Reid, O'Donnell, and Labavitch, unpublished).
4. When we silenced petunia ACC oxidase (ACO; the terminal step in ethylene biosynthesis) we used vector combining a fragment from ACO4 (GenBank accession number L21979) and chalcone synthase (CHS). The silenced (white) flowers produced much less ethylene and lasted longer than the violet control flowers. In petunia, ACO is encoded by a multigene family, and we showed that the reduced ethylene production of the silenced sectors was correlated with a reduction not only in the abundance of ACO4 transcripts (to 5% of the controls) but also in that of ACO1 transcripts (to 23% of the controls). Co-silencing of two genes (PCNA and CH42) in one TRV vector has also recently been demonstrated in *Jatropha* (14).
5. In initial studies, we used ESTs obtained from the petunia floral EST database from the University of Florida, excising inserts from the plasmids and cloning them directly into the MCS of the TRV RNA2 vector. Puzzled by the low level of silencing obtained with these constructs, we eventually deduced that the polyA tail on the EST clones interfered with the silencing mechanism (Chen, Jiang, and Reid, unpublished). The problem was overcome by interposing an amplification step, using primers specific to each EST.

6. To generate cDNA, total RNA should be extracted from mixed tissues of leaves, flowers, and roots to increase the probability of amplification of genes that express only in specific tissues.
7. The orientation of the fragments in the RNA2 vector does not seem to affect the efficacy of silencing.
8. Acetosyringone is an activator of *Vir* genes and improves the transformation efficiency by *Agrobacterium*.
9. In petunia, tomato, and tobacco, we introduce *Agrobacterium* by syringe infiltration of the leaves (**Fig. 6.2**). Another technique that warrants testing involves dripping the mixed inoculum onto the meristem and then gently pinching, a technique reported to be effective with California poppy (*Eschscholzia*) (34). Using an artist airbrush attached to a pressure compressor to spray the plants with the cell mixtures has been successfully tested for VIGS in tomato (5). An alternative technique that we have tried with success is vacuum or pressure infiltration of seedlings (ca. 1–2 weeks old) (14, 19). Other researchers have been successful in inoculating older tissues, including developing fruits, by direct injection of the cell mixtures (6, 35).
10. Optimal efficiency is obtained when plants are young (two to four true leaves) and growing rapidly.
11. In our initial VIGS studies, we found that growth conditions were critical for efficient silencing. Temperature seems to have profound effects on vector DNA accumulation and the spread of silencing signals. Plants grown under the relatively variable conditions of a standard greenhouse showed variable silencing; we found it essential to provide a day/night temperature variation. The need for a day/night temperature variation for successful systemic infection has also been reported in tospovirus infection of capsicum and *N. benthamiana* (36). These authors suggested that long-distance transport of the virus may be inhibited at high temperature. Fu et al. (37) reported that gene silencing with TRV in tomato was enhanced by low temperature and low humidity, but Szittya et al. (38) found that low temperatures inhibited silencing by inhibiting siRNA generation. These apparently conflicting findings may reflect a requirement for low temperature to allow initial virus replication and systemic movement and a requirement for higher temperatures to activate the silencing mechanism sufficiently to silence endogenous gene expression. Recently, Tuttle et al. (29) reported that Gemini virus-mediated VIGS is enhanced by low temperature in cotton. Cotton plants grown at high temperature (30/26°C) had reduced

viral DNA accumulation compared to the plants grown at low temperature (22/18°C). Efficiency of gene silencing also decreased when plants were grown at high temperature. Low temperature enhanced gene silencing throughout the life of cotton plants, although onset of silencing was delayed substantially.

12. Early studies used the silencing of PDS to show the effect of VIGS; infected plants show characteristic photobleaching symptoms resulting from the inhibition of biosynthesis of protective carotene. While silencing of PDS serves as a clear reporter, the phenotype has obvious disadvantages, as the photobleaching is concomitant with destruction of the photosynthetic apparatus. Less detrimental reporters for silencing in green tissues would be useful. One candidate is silencing of green fluorescent protein (GFP) in plants that are constitutively expressing this transgene (39). An anthocyanin-based reporter system has been demonstrated in tomato (19, 35). In petunia, the PDS photobleaching phenotype can be seen from 7 days after inoculation; the CHS phenotype is seen in flower petals as soon as they open. Systemic spread of silencing into developing leaves and flowers seems to be dependent on growth conditions.

Acknowledgments

We appreciate the assistance of Dr. Dinesh-Kumar, Yale University, in providing the TRV constructs. Goldsmith Seeds generously donated seeds of blue-flowered hybrid petunia cultivars. Our studies have partially been supported by grants from the American Floral Endowment, BARD (the United States–Israel Binational Agricultural Research and Development Fund, Research Grant No. IS-3815-05) and the USDA Floriculture Initiative.

References

1. Napoli, C., Lemieux, C., and Jorgensen, R. (1990) Introduction of a chimeric chalcone synthase gene into petunia results in reversible co-suppression of homologous genes *in trans*. *Plant Cell* **2**, 279–290.
2. Voinnet, O. (2001) RNA silencing as a plant immune system against viruses. *Trends Genet.* **17**, 449–459.
3. Kumagai, M. H., Donson, J., Della-Ciopa, G., Harvey, D., Hanley, K., and Grill, L. K. (1995) Cytoplasmic inhibition of carotenoid biosynthesis with virus-derived RNA. *Proc. Natl. Acad. Sci. USA* **92**, 1679–1683.
4. Turnage, M. A., Muangsan, N., Peele, C. G., and Robertson, D. (2002) Geminivirus-based vectors for gene silencing in Arabidopsis. *Plant J.* **30**, 107–114.
5. Liu, Y., Schiff, M., and Dinesh-Kumar, S. P. (2002) Virus-induced gene silencing in tomato. *Plant J.* **31**, 777–786.

6. Fu, D. Q., Zhu, B. Z., Zhu, H. L., Jiang, W. B., and Luo, Y. B. (2005) Virus-induced gene silencing in tomato fruit. *Plant J.* **43**, 299–308.
7. Faivre-Rampant, O., Gilroy, E. M., Hrubikova, K., Hein, I., Millam, S., Loake, G. J., Birch, P., Taylor, M., and Lacomme, C. (2004) Potato virus X-induced gene silencing in leaves and tubers of potato. *Plant Physiol.* **134**, 1308–1316.
8. Holzberg, S., Brosio, P., Gross, C., and Pogue, G. P. (2002) Barley stripe mosaic virus-induced gene silencing in a monocot plant. *Plant J.* **30**, 315–327.
9. Lacomme, C., Hrubikova, K., and Hein, I. (2003) Enhancement of virus-induced gene silencing through viral-based production of inverted-repeats. *Plant J.* **34**, 543–553.
10. Tai, Y.-S., Bragg, J. N., and Edwards, M. C. (2005) Virus vector for gene silencing in wheat. *Biotechniques* **39**, 310–314.
11. Scofield, S. R., Huang, L., Brandt, A. S., and Gill, B. S. (2005) Development of a virus-induced gene-silencing system for hexaploid wheat and its use in functional analysis of the Lr21-mediated leaf rust resistance pathway. *Plant Physiol.* **138**, 2165–2173.
12. Scofield, S. R. and Nelson, R. S. (2009) Resources for virus-induced gene silencing in the grasses. *Plant Physiol.* **149**, 152–157.
13. Ding, X. S., Schneider, W. L., Chaluvadi, S. R., Mian, M. A. R., and Nelson, R. S. (2006) Characterization of a Brome mosaic virus strain and its use as a vector for gene silencing in monocotyledonous hosts. *Mol. Plant Microbe Interact.* **19**, 1229–1239.
14. Ye, J., Qu, J., Bui, H. T. N., and Chua, N.-H. (2009) Rapid analysis of *Jatropha curcas* gene functions by virus-induced gene silencing. *Plant Biotechnol. J.* 7, 964–976.
15. Nagamatsu, A., Masuta, C., Senda, M., Matsuura, H., Kasai, A., Hong, J.-S., Kitamura, K., Abe, J., and Kanazawa, A. (2007) Functional analysis of soybean genes involved in flavonoid biosynthesis by virus-induced gene silencing. *Plant Biotechnol. J.* **5**, 778–790.
16. Igarashi, A., Yamagata, K., Sugai, T., Takahashi, Y., Sugawara, E., Tamura, A., Yaegashi, H., Yamagishi, N., Takahashi, T., Isogai, M., Takahashi, H., and Yoshikawa, N. (2009) Apple latent spherical virus vectors for reliable and effective virus-induced gene silencing among a broad range of plants including tobacco, tomato, *Arabidopsis thaliana*, cucurbits, and legumes. *Virology* **386**, 407–416.
17. Yamagishi, N. and Yoshikawa, N. (2009) Virus-induced gene silencing in soybean seeds and the emergence stage of soybean plants with Apple latent spherical virus vectors. *Plant Mol. Biol.* **71**, 15–24.
18. Chen, J. C., Jiang, C. Z., Gookin, T. E., Hunter, D. A., Clark, D. G., and Reid, M. S. (2004) Chalcone synthase as a reporter in virus-induced gene silencing studies of flower senescence. *Plant Mol. Biol.* **55**, 521–530.
19. Jiang, C. Z., Lu, F., Imsabai, W., Meir, S., and Reid, M. S. (2008) Silencing polygalacturonase expression inhibits tomato petiole abscission. *J. Exp. Bot.* **59**, 973–979.
20. Reid, M., Chen, J.-C., and Jiang, C.-Z. (2009) Virus-induced gene silencing for functional characterization of genes in petunia. In: *Petunia* (eds. Gerats, T. and Strommer, J.), Springer, New York, pp. 381–394.
21. English, J. J., Mueller, E., and Baulcombe, D. C. (1996) Suppression of virus accumulation in transgenic plants exhibiting silencing of nuclear genes. *Plant Cell* **8**, 179–188.
22. Angell, S. M. and Baulcombe, D. C. (1997) Consistent gene silencing in transgenic plants expressing a replicating potato virus X RNA. *EMBO J.* **16**, 3675–3684.
23. Lu, R., Martin-Hernandez, A. M., Peart, J. R., Malcuit, I., and Baulcombe, D. C. (2003) Virus-induced gene silencing in plants. *Methods* **30**, 296–303.
24. Lu, R., Malcuit, I., Moffett, P., Ruiz, M. T., Peart, J., Wu, A. J., Rathjen, J. P., Bendahmane, A., Day, L., and Baulcombe, D. C. (2003) High throughput virus-induced gene silencing implicates heat shock protein 90 in plant disease resistance. *EMBO J.* **22**, 5690–5699.
25. Kjemtrup, S., Sampson, K. S., Peele, C. G., Nguyen, L. V., Conkling, M. A., Thompson, W. F., and Robertson, D. (1998) Gene silencing from plant DNA carried by a Geminivirus. *Plant J.* **14**, 91–100.
26. Ratcliff, F., Martin-Hernendez, A. M., and Baulcombe, C. D. (2001) Tobacco rattle virus as a vector for analysis of gene function by silencing. *Plant J.* **25**, 237–245.
27. Tao, X. and Zhou, X. (2004) A modified viral satellite DNA that suppresses gene expression in plants. *Plant J.* **38**, 850–860.
28. Fofana, I., Sangar, A., Collier, R., Taylor, C., and Fauquet, C. (2004) A geminivirus-induced gene silencing system for gene function validation in cassava. *Plant Mol. Biol.* **56**, 613–624.
29. Tuttle, J. R., Idris, A. M., Brown, J. K., Haigler, C. H., and Robertson, D. (2008) Geminivirus-mediated gene silencing from cotton leaf crumple virus is enhanced by low temperature in cotton. *Plant Physiol.* **148**, 41–50.

30. Constantin, G. D., Krath, B. N., MacFarlane, S. A., Nicolaisen, M., Johansen, I. E., and Lund, O. S. (2004) Virus-induced gene silencing as a tool for functional genomics in a legume species. *Plant J.* **40**, 622–631.
31. Zhang, C. and Ghabrial, S. A. (2006) Development of Bean pod mottle virus-based vectors for stable protein expression and sequence-specific virus-induced gene silencing in soybean. *Virology* **344**, 401–411.
32. Liu, Y., Schiff, M., Marathe, R., and Dinesh-Kumar, S. P. (2002) Tobacco *Rar1*, *EDS1* and *NPR1/NIM1* like genes are required for *N*-mediated resistance to tobacco mosaic virus. *Plant J.* **30**, 415–429.
33. Gould, B. and Kramer, E. M. (2007) Virus-induced gene silencing as a tool for functional analyses in the emerging model plant *Aquilegia* (columbine, *Ranunculaceae*). *Plant Methods* **3**, 6.
34. Wege, S., Scholz, A., Gleissberg, S., and Becker, A. (2007) Highly efficient virus-induced gene silencing (VIGS) in California poppy (*Eschscholzia californica*): an evaluation of VIGS as a strategy to obtain functional data from non-model plants. *Ann. Bot. (Lond.)* **100**, 641–649.
35. Orzaez, D., Medina, A., Torre, S., Fernandez-Moreno, J. P., Rambla, J. L., Fernandez-del-Carmen, A., Butelli, E., Martin, C., and Granell, A. (2009) A visual reporter system for virus-induced gene silencing in tomato fruit based on anthocyanin accumulation. *Plant Physiol.* **150**, 1122–1134.
36. Roggero, P., Dellavalle, G., Ciuffo, M., and Pennazio, S. (1999) Effects of temperature on infection in *Capsicum* sp. and *Nicotiana benthamiana* by impatiens necrotic spot tospovirus. *Eur. J. Plant Pathol.* **105**, 509–512.
37. Fu, D. Q., Zhu, B. Z., Zhu, H. L., Zhang, H. X., Xie, Y. H., Jiang, W. B., Zhao, X. D., and Luo, K. B. (2006) Enhancement of virus-induced gene silencing in tomato by low temperature and low humidity. *Mol. Cells* **21**, 153–160.
38. Szittya, G., Silhavy, D., Molnar, A., Havelda, Z., Lovas, A., Lakatos, L., Banfalvi, Z., and Burgyan, J. (2003) Low temperature inhibits RNA silencing-mediated defence by the control of siRNA generation. *EMBO J.* **22**, 633–640.
39. Burch-Smith, T. M., Schiff, M., Liu, Y., and Dinesh-Kumar, S. P. (2006) Efficient virus-induced gene silencing in Arabidopsis. *Plant Physiol.* **142**, 21–27.

Chapter 7

Local RNA Silencing Mediated by Agroinfiltration

Jutta Maria Helm, Elena Dadami, and Kriton Kalantidis

Abstract

Agroinfiltration is a very fast and powerful method to express *in planta* any sequences in a transient fashion. Agroinfiltration has proven very useful for the overexpression of proteins in the infiltrated zone when a short-term effect can be informative. However, it has been a real success story in the induction of local and eventually systemic silencing. Here, we describe the use of agroinfiltration for the induction of local silencing of an endogene or a transgene, for the systemic silencing of a transgene and for co-infiltration assays. We also provide protocols for the evaluation of the efficiency of the assay, by detecting the specific siRNAs characteristic of RNA silencing and measuring the effects on the target sequences.

Key words: Agroinfiltration, co-infiltration, patch assay, PTGS, RNAi, RNA silencing, transient expression.

1. Introduction

Higher eukaryotes have developed a mechanism of sequence-specific RNA degradation called RNA silencing. The central part of the RNA degradation pathway is the generation of short interfering RNA (siRNA) from double-stranded RNA by a RNase III-type nuclease, Dicer. The siRNAs are incorporated into the RNA-induced silencing complex (RISC), and after strand separation the remaining single-stranded RNA guides the sequence-specific cleavage of a target RNA (1).

This pathway characterizes the response of cells to exogenous sequences (viruses, transgenes, etc.), although it can also be directed to target endogenous sequences. Under physiological conditions, other small RNA molecules function to silence endogenous sequences (2). Posttranscriptional gene silencing

H. Kodama, A. Komamine (eds.), *RNAi and Plant Gene Function Analysis*, Methods in Molecular Biology 744,
DOI 10.1007/978-1-61779-123-9_7, © Springer Science+Business Media, LLC 2011

(PTGS) in plants is highly conserved and considered as a plant-immune response. Its main biological functions are to assist in the resistance to viruses and other exogenous RNAs (for example, viroids or RNA from agrobacterial T-DNA-encoded genes), secure the stability of the genome through the suppression of transposon activity and provide an additional mechanism for the regulation of endogenes (3, 4).

The inhibition of nopaline synthase in tobacco after expression of its corresponding asRNA (5) was the first report of RNA silencing pathways in plants. The phenomenon of co-suppression describes the downregulation of endogenous sequences by strong overexpression of the same sequence transcribed from a transgene. It was discovered by two groups trying to increase the purple pigmentation of petunia by overexpression of chalcone synthase (CHS), an enzyme of the anthocyanin pathway, which lead to the formation of white petals (6–8).

In plants, cytoplasmic RNA silencing can be induced efficiently by agroinfiltration (9), a method for transiently expressing high levels of transgenes. *Agrobacterium tumefaciens* is used as a vector for inserting its T-DNA into plant cell nuclei where they can be transcribed. It is a rapid, versatile and convenient way for achieving a very high level of gene expression in a distinct and defined zone of a leaf, and can be used not only for inducing silencing processes (10, 11) and dissecting gene functions but also for simple protein production (12) and purification (11, 13–15).

Transient gene expression systems offer a number of advantages compared to the use of transgenic plants. It is fast and easy to accomplish and can therefore be used to test the functionality of a construct before it is used for the laborious process of transgenic plant production. It can also be used for plants which are not easy to regenerate in tissue culture. Transient delivery avoids any positional effects that could bias the gene expression (16). The expression levels are usually very high and result in a large yield of the desired product. In addition the expression of the transgene is induced at an advanced developmental stage, therefore effects of potentially detrimental constructs can be also assayed (12).

Nevertheless, agroinfiltration assays have some limitations as well. They are influenced by the physiological condition of the plants and experimental variables which affect the virulence of the Agrobacteria. These factors can lead to necrosis and fast leaf wilting before any effects can be detected (17).

Another useful application of agroinfiltration is co-infiltration. It is a quick and reliable tool to analyse the interaction of different transgenes and their transcription and/or translation products in a plant cell. These interactions could be the following:

(1) RNA/RNA, in the case of analysing the *in trans* capacity of a hairpin construct (and its Dicer by-products) to induce PTGS to a sensor construct (18);

(2) RNA/DNA, in the case of analysing the *in trans* capacity of a hairpin construct to induce *in trans* RdDM to a sensor construct (19);

(3) RNA/protein, in the case of analysing the *in trans* capacity of a hairpin construct (and its Dicer by-products) that leads to translational arrest of a sensor construct or identification of novel silencing suppressors that bind/stabilize/degrade RNAi-inducing molecules (20);

(4) Protein/protein, in the case of identifying, as above, novel silencing suppressors that interact with vital enzymes of the RNAi pathway (21).

2. Materials

All chemicals unless mentioned otherwise were purchased from Merck KGaA, Darmstadt, Germany, or Sigma-Aldrich, St. Louis, MO.

2.1. Agroinfiltration

1. Lysogeny broth (LB): 10 g/L tryptone, 5 g/L yeast extract, 10 g/L NaCl, pH 7.0, with 100 μg/mL rifampicin (Duchefa Biochemie B.V., Haarlem, Netherlands) and respective vector resistance antibiotic.
2. MMA solution: MS salts (Duchefa Biochemie B.V., Haarlem, Netherlands), 10 mM 2-(*N*-morpholino) ethanesulphonic acid (MES), pH 5.6, 200 μM acetosyringone added prior to use from a 200 mM stock (store at –20°C).
3. 10 mM $MgCl_2$.
4. 1-mL Syringe (HMD Healthcare Ltd, Horsham, UK).

2.2. Visualization of GFP

1. Handheld UV lamp UVP Black-Ray® B 100AP/R.

2.3. Preparation of Agarose–Formaldehyde Gels for Northern Analysis of mRNAs

1. 10× MOPS stock solution: 200 mM 3-(*N*-morpholino) propanesulphonic acid (MOPS), 50 mM sodium acetate, 10 mM EDTA, pH 7.0 with NaOH.
2. Agarose, 37% formaldehyde, 2 μg/μL ethidium bromide.
3. 5× RNA loading dye: 0.16% (v/v) saturated aqueous bromophenol blue solution, 36 mM EDTA (pH 8.0), 2.6% (v/v) (0.89 M) formaldehyde, 20% (v/v) autoclaved glycerol, 31% (v/v) formamide, 4% (v/v) MOPS; final volume is adjusted with autoclaved distilled water.

2.4. Preparation of Denaturing Polyacrylamide Gel Electrophoresis (PAGE) for Northern Analysis of Small RNAs

1. 5× TBE: 0.45 M Tris base, 0.45 M boric acid, 10 mM EDTA, pH 8.0.
2. 40% Acrylamide/bis solution (38:2): 380 g acrylamide and 20 g *N,N′*-methylenediacrylamide.
3. 10% Ammonium persulphate (APS): Prepare solution and immediately freeze in 1 mL aliquots, store at –20°C.
4. Urea, *N,N,N′,N′*-Tetramethylethylenediamine (TEMED).
5. 2× PAGE loading buffer: 98% (v/v) deionized formamide, 10 mM EDTA, 0.025% (w/v) xylene cyanol FF and 0.025% (w/v) bromophenol blue, final volume is adjusted with autoclaved distilled water.

2.5. Blotting, Membranes and Cross-linking

1. 20× SSC: 3 M sodium chloride, 0.3 M sodium citrate, pH 7.0.
2. Nytran N membrane (GE Healthcare, Little Chalfont, UK): Pore size 0.45 μm for agarose gels and 0.2 μm for polyacrylamide gels.
3. Cross-linking: Stratagene Stratalinker (Agilent, La Jolla, CA).
4. Blotting: SD20 Semi Dry Midi (Cleaver Scientific Ltd, Warwickshire, UK).

2.6. Hybridization and Washes

1. 50× Denhardt's: 1% (w/v) Ficoll 400, 1% (w/v) polyvinylpyrrolidone (PVP), 1% (w/v) bovine serum albumin (BSA).
2. Pre-hybridization for detection of large RNAs: 5× SSC, 1× Denhardt's solution, 1% (w/v) SDS and 0.25 mg/mL tRNA carrier, carried out at around 65°C (for 100% homology of probe) for 1 h. Hybridization solution includes labelled DNA probe at approximately 10^6 cpm/mL, carried out at 65°C overnight.
3. Pre-hybridization for detection of small RNAs: 5× SSC, 20 mM Na_2HPO_4, 7% (w/v) SDS, 1× Denhardt's solution, carried out at 50°C for 1 h. Hybridization solution includes labelled DNA probe at approximately 10^6 cpm/mL, carried out at 50°C overnight.
4. 2.5× RadPrime buffer: 125 mM Tris–HCl, pH 6.8, 12.5 mM $MgCl_2$, 25 mM β-mercaptoethanol.
5. 3 μg/μL Random primers (Invitrogen, Carlsbad, CA).
6. 3,000 Ci/mmol [α^{32}P]dATP and [α^{32}P]dCTP (PerkinElmer, Waltham, MA).
7. 0.5 M dGTP, 0.5 M dTTP (Bioron GmbH, Ludwigshafen, Germany).

8. DNA polymerase I Klenow fragment 5 U/μL (Minotech, Heraklion, Greece).
9. Illustra MicroSpin S-200 column (GE Healthcare, Little Chalfont, UK).
10. Wash I: 2× SSC and 0.3% SDS; wash II: 1× SSC and 0.3% SDS.
11. X-Ray films Fuji Super RX (Fujifilm, Minato Tokyo, Japan).

3. Methods

Prior to assaying RNA silencing by agroinfiltration, a hairpin construct needs to be prepared and cloned into a binary vector (listed in 22).

Standard hairpins contain the desired sequence in an inverted repeat separated by a spacer sequence. Intron sequences are usually used since they have been reported to enhance the efficiency of silencing (23). The sequences can be inserted into a vector by using different primers with restriction sites and subsequent ligation steps. Another method is the Gateway® technology (Invitrogen, Carlsbad, CA), where the desired sequence is inserted into a vector by recombination (methods described in 24).

Agrobacteria can be transformed using three different methods: electroporation, freeze/thawing, and triparental mating (methods described in 25).

3.1. Agroinfiltration

This method is based on Schöb et al (26):

1. Agrobacterial strains should be stored according to the facilities as a stab or a cryoculture to guarantee long-time storage survival.
2. LB liquid cultures (5 mL; *see* **Note 1**) containing the appropriate antibiotics are prepared and inoculated with the stored strain in a sterile environment.
3. The cells are grown at 28°C until an OD_{600} between 0.1 and 0.8 is reached (*see* **Note 2**).
4. The Agrobacteria are pelleted by centrifugation at 2,500×*g* at 4°C and the supernatant is discarded and to be treated as biological waste.
5. The cells are resuspended in 5 mL of MMA (*see* **Note 3**) and incubated in a shaker at 28°C for at least 2 h.
6. The bacteria are pelleted by centrifugation as above and the resulting pellet is resuspended in 2 mL of 10 mM $MgCl_2$ (*see* **Note 4**). This washing step is repeated twice.

7. The OD_{600} of the bacteria is adjusted to approximately 0.25 (*see* **Note 5**).
8. The solution is sucked into a 1-mL syringe.
9. The *Nicotiana* sp. leaves (*see* **Notes 6** and 7) are punctured on their lower surface with a needle.
10. The tip of the syringe is pressed gently on the small hole and a finger is placed on the other side of the leaf for stabilization.
11. The plunger is pressed gently and by applying constant slight pressure, the liquid can be observed entering the leaf.
12. To infiltrate a whole leaf, steps 8–11 can be applied repeatedly.
13. **Figure 7.1** shows a graphic display of steps 9–11.

Alternatively, leaves can be infiltrated by vacuum; a sample protocol can be found elsewhere (16).

Fig. 7.1. Agroinfiltration in *Nicotiana benthamiana*. **a** Using a needle, a small tear on the bottom of the leaf is made. **b** The agroinfiltration solution is delivered in the leaf with a syringe. **c** The agroinfiltration spot on the leaf is indicated.

3.2. Observation of Visible Local Silencing

The observation of local silencing depends on the target of the infiltrated hairpin. Transgenes containing visible markers such as GFP can be easily observed under UV light, usually the decrease of the green fluorescence is detectable locally 3–4 days post-infiltration and systemically 1–2 weeks later.

3.3. Northern Analysis for Detecting the Levels of the Targeted RNA

Since an infiltrated hairpin will target an mRNA, the detection of RNA is the most straightforward method:

1. RNA is extracted by the common laboratory method.
2. The infiltrated hairpin can then be detected by separating 10–20 μg of total RNA on denaturing agarose/formaldehyde gels. The gel casting device and the buffer chamber are washed thoroughly with liquid soap, water and ethanol. Agarose (1.2 g) in 88 g H_2O is melted in a microwave, slightly cooled and evaporated water is replenished. 10X MOPS (1 mL), 7 μL ethidium bromide

and 1.8 mL of 37% formaldehyde (final concentration is 0.7%, v/v) are added, mixed and the gel is cast.

3. After polymerization, the gel is pre-run at 100 V in 1× MOPS with 0.7% (v/v) formaldehyde buffer for some minutes. The samples in 1× RNA loading buffer are boiled, quick chilled and loaded onto the gel. A constant voltage of 100 V is applied until the RNA has fractionated sufficiently.
4. The gel is photographed, rinsed in water and used for capillary blotting in 10× SSC overnight. The membrane is dried, UV cross-linked with 120,000 μJ/cm^2 and used for hybridization (see below).
5. Resulting small RNAs from the hairpin and the downregulated target can be detected by separating 20–50 μg of total RNA on 15% denaturing polyacrylamide gels: If no precast gels are available, glass plates are washed thoroughly with liquid soap, water and ethanol. Acrylamide/bis mix (22.5 mL), 28.3 g urea, 12 mL of 5× TBE in 60 mL total volume are incubated at 42°C until the urea is dissolved (*see* **Note 8**). The mixture is polymerized with 225 μL APS and 45 μL TEMED (*see* **Note 9**) and cast in the assembled glass plates.
6. The polymerized gels are pre-run at 22 W until they reach a temperature of 50°C.
7. The samples are boiled in 1× PAGE loading buffer and quick chilled on ice. Prior to loading, the wells must be rinsed of released urea with 1× TBE, which is pressed into them with a syringe (*see* **Note 10**). A constant low wattage of 11 W is applied until the samples have completely entered the gel. Then the wattage can be increased, taking care that the temperature of the gel does not exceed 50°C. The gels are run until the bromophenol blue front, which at this gel thickness runs at approximately 10 nt, reaches the lower margin.
8. The gels are rinsed in 1× TBE and the RNA is blotted with a SD20 Semi Dry Midi device, using two times 5 gel-size Whatman paper soaked in 1× TBE surrounding the equilibrated gel and membrane. A constant amperage of 3 mA/cm^2 membrane area is applied for 30 min at 4°C and afterwards the RNA is cross-linked on the membrane by UV radiation of 120,000 μJ/cm^2.
9. The membranes are pre-hybridized with 7–15 mL hybridization buffer (*see* **Note 11**) for 1 h at 50–65°C. Purified DNA template (100 ng) is denatured by boiling for 3 min and subsequent quick chilling. The typical reaction volume is 50 μL, consisting of 20 μL RadPrime buffer, 1 μL random primers, 1 μL dTTP, 1 μL dGTP,

20 μCi [α^{32}P]dATP and 20 μCi [α^{32}P]dCTP, as well as 20 U Klenow fragment. After an incubation of 60 min at 37°C, the labelled probe is purified from unincorporated nucleotides using a MicroSpin S-200 column (GE Healthcare, Little Chalfont, UK) according to the manufacturer's protocol. Prior to use, the probe is denatured by boiling for 3 min and subsequent quick chilling. The probe is added to the pre-hybridized membrane and it is hybridized overnight at 50–65°C.

10. Next day the membrane is rinsed at room temperature with wash I and then washed twice at hybridization temperature for 30 min with excess wash I (*see* **Note 12**). When large RNAs are to be detected, another washing step is performed for 10 min at hybridization temperature with wash II. The membrane is rinsed with 2× SSC, sealed in plastic bags and subsequently exposed on X-ray films. Signals can be detected after 1–10 day exposure at –80°C depending on the amount of the targeted RNA.

An example of northern hybridizations for siRNAs, hairpin RNA and target mRNA sequences using the protocols described above is shown in **Fig. 7.2**.

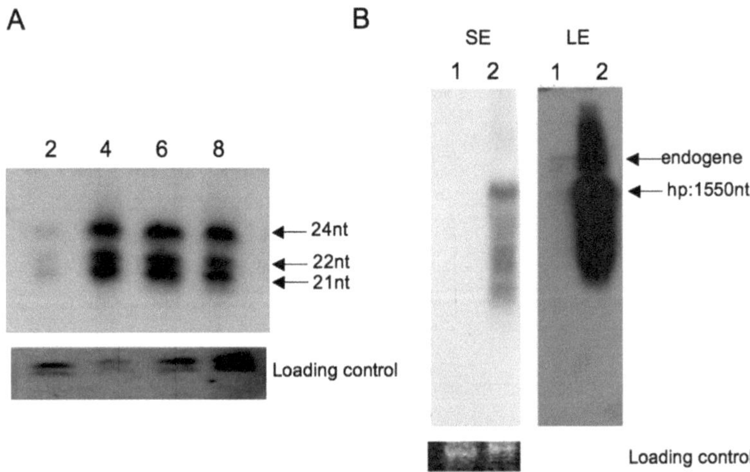

Fig. 7.2. Detection of induced silencing by northern blotting and hybridization. **a** Detection of GFP siRNAs in 16c *N. benthamiana* line following agroinfiltration of a silencing trigger (18). Accumulation of GFP siRNAs following agroinfiltration of a GFP–hairpin expressing construct (hpGFP S-AS). siRNA detection 2, 4, 6 and 8 days post-infiltration. Note that all three classes of siRNAs (21 nt, 22 nt and 24 nt long) can be clearly detected. U1 RNA detection provides loading control. **b** Silencing of an endogene using a homologous hairpin construct in *N. benthamiana*. Detection of the hairpin and degradation products (*lane 2*) 3 days post-agroinfiltration. In the longer exposure (LE) of the same membrane, the targeted endogene can also be detected in the non-agroinfiltrated sample (LE, *lane 1*). The *arrow* indicates the size of the hairpin. SE, short exposure; LE, long exposure. Ethidium bromide-stained rRNA was used as loading control (Dadami, unpublished results).

3.4. Other Methods for Detecting Target Downregulation

1. qPCR for determining transcript target levels: The downregulation of the targeted mRNA can be determined by quantitative RT-PCR. For this purpose, 1–2 μg of total DNase-treated total RNA is used for RT-PCR. The resulting cDNA is diluted 1:10 to prevent inhibitory effects in the further steps. The qPCR is executed depending on the laboratory facilities. The cDNAs are analysed in triplicates; in addition, a standard curve of known amounts of the gene of interest as well as housekeeping genes and internal standards has to be prepared at least in duplicates. The degree of downregulation of the target mRNA can be determined with appropriate software. Detailed protocols for qPCR can be found elsewhere.
2. Western blot to detect downregulation of protein level: If a suitable antibody is available, the downregulation might also be detectable at the protein level, in case the targeted protein is not too stable.

3.5. Observation of Systemic Spread

If the target of the infiltrated hairpin is a transgene, the silencing is able to spread systemically out of the infiltration zone and can eventually lead to transgene silencing in the whole plant. The same methods as above can be used for detecting the silencing in leaves different from the infiltration site. **Figure 7.3** shows silencing of a *GFP* transgene induced by agroinfiltration and the manifestation of systemic spread.

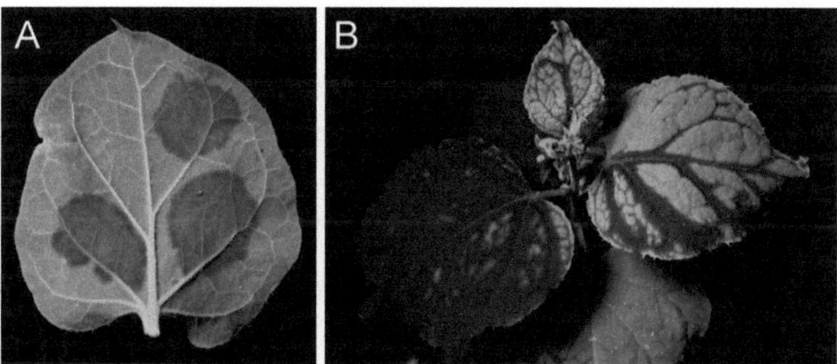

Fig. 7.3. Manifestations of RNA silencing after agroinfiltration. **a** Silencing of the GFP transgene of *N. benthamiana* line 16c after agroinfiltration of a GFP hairpin construct. **b** Systemic spread of the GFP silencing out of the infiltration zone leads to a completely silenced plant. Reproduced with permission from (30) © the Biochemical Society and must include a link to the journal Web site (http://www.biolcell.org).

4. Notes

1. Depending on the required amount of leaves the cultures size can be upscaled; 5 mL will be usually sufficient for at least five leaves of 25 cm^2.

2. Depending on the amount of inoculum, the growth of the cultures can take 2 days and more since Agrobacteria divide approximately every 2 h.
3. The incubation with acetosyringone makes the bacteria more virulent and thus increases the transformation rate (16).
4. The volume for resuspension depends on the initial culture volume and should be increased accordingly.
5. Higher ODs can easily lead to necrosis of the infiltrated areas; too low ODs might result in poor expression of the infiltrated genes. In general, the optimal OD_{600} is dependent on the *A. tumefaciens* strain and should be determined empirically.
6. This method can be applied for almost any leaf tissue (summary in **Table 7.1**).
7. The physiological stage of the plants is essential for the success of the infiltration. Young, healthy plants are preferable; in addition, the watering should be stopped up to 1 day before infiltration, otherwise the leaves might not take up the infiltrated liquid.

Table 7.1
Agroinfiltration parameters in different plant species

Organism	Method	*Agrobacterium* strain	Optimal OD_{600}	Highest expression	Reference
Arabidopsis[a]	Agroinfiltration	C58C1	No data	Rosette leaves = ~30	(17)
Nicotiana sp.[a]	Agroinfiltration	C58C1, GV3101	0.1–0.8	Basal region of leaf	(17, 18)
Tomato[a]	Agroinfiltration	1D1249 (no necrosis), C58C1	0.1–0.8	Basal region of leaf	(17)
Lettuce[a]	Agroinfiltration	C58C1	0.3–0.8	Basal region of leaf	(17)
Potato	Agroinfiltration	GV3101, LBA4404	0.2–0.5	Terminal leaflets, 5–6-week-old plants	(27)
Strawberry	Agroinjection	AGL0	~0.8	fruit	(28)
Grapes	Agroinfiltration in tissue culture	AGL1, LBA4404	~0.2	Vacuum-infiltrated whole leaves	(29)

[a]In these species *A. tumefaciens* strain C58C1 resulted in higher transgene expression; in tomato, this strain resulted in a necrotic phenotype several days post-infiltration.

8. The mix can also be prepared in the evening; the urea will dissolve overnight at room temperature.
9. The amounts of TEMED and 10% APS vary to a great extent in the literature. In our hands also, very different amounts are used and work.
10. This step is essential since traces of released urea will have a detrimental effect on equal loading and can lead to loss of samples.
11. The amount of hybridization buffer depends on the size of the membrane. Usually the smallest possible cylinder volume is used for hybridization and the membrane should be covered entirely during hybridization. In addition, prehybridization can be carried out in a larger volume, but during hybridization the probe should be diluted in the smallest possible volume (totally covered membrane).
12. The volume of the washing solutions should be at least twice the hybridization buffer volume.

Acknowledgements

Jutta Maria Helm has been supported by the European Union through the Marie Curie Fellowship MEST-CT-2004-007295 FAMED and Elena Dadami by the University of Crete (ELKE medium size grant) and IKY (IKYDA gr 126).

References

1. Baulcombe, D. C. (2006) Short silencing RNA: the dark matter of genetics? *Cold Spring Harb. Symp. Quant. Biol.* **71**, 13–20.
2. Ghildiyal, M. and Zamore, P. D. (2009) Small silencing RNAs: an expanding universe. *Nat. Rev. Genet.* **10**, 94–108.
3. Vaucheret, H., Béclin, C., and Fagard, M. (2001) Post-transcriptional gene silencing in plants. *J. Cell. Sci.* **114**, 3083–3091.
4. Baulcombe, D. (2004) RNA silencing in plants. *Nature* **431**, 356–363.
5. Rothstein, S. J., Dimaio, J., Strand, M., and Rice, D. (1987) Stable and heritable inhibition of the expression of nopaline synthase in tobacco expressing antisense RNA. *Proc. Natl. Acad. Sci. USA* **84**, 8439–8443.
6. Napoli, C., Lemieux, C., and Jorgensen, R. (1990) Introduction of a chimeric chalcone synthase gene into *Petunia* results in reversible co-suppression of homologous genes *in trans*. *Plant Cell* **2**, 279–289.
7. van der Krol, A. R., Mur, L. A., Beld, M., Mol, J. N., and Stuitje, A. R. (1990) Flavonoid genes in *Petunia*: addition of a limited number of gene copies may lead to a suppression of gene expression. *Plant Cell* **2**, 291–299.
8. Benedito, V. A., Visser, P. B., Angenent, G. C., and Krens, F. A. (2004) The potential of virus-induced gene silencing for speeding up functional characterization of plant genes. *Genet. Mol. Res.* **3**, 323–341.
9. Vaucheret, H. (1994) Promoter-dependent trans-inactivation in transgenic tobacco

plants: kinetic aspects of gene silencing and gene reactivation. *CR. Acad. Sci. Paris Ser. III* **317**, 310–323.

10. Voinnet, O. and Baulcombe, D. C. (1997) Systemic signalling in gene silencing. *Nature* **389**, 553.
11. Johansen, L. K. and Carrington, J. C. (2001) Silencing on the spot. Induction and suppression of RNA silencing in the *Agrobacterium*-mediated transient expression system. *Plant Physiol.* **126**, 930–938.
12. Fischer, R., Vaquero-Martin, C., Sack, M., Drossard, J., Emans, N., and Commandeur, U. (1999) Towards molecular farming in the future: transient protein expression in plants. *Biotechnol. Appl. Biochem.* **30**, 113–116.
13. Mlotshwa, S., Voinnet, O., Mette, M. F., Matzke, M., Vaucheret, H., Ding, S. W., Pruss, G., and Vance, V. B. (2002) RNA silencing and the mobile silencing signal. *Plant Cell* **14**, S289–S301.
14. Tenllado, F., Barajas, D., Vargas, M., Atencio, F. A., Gonzalez-Jara, P., and Diaz-Ruiz, J. R. (2003) Transient expression of homologous hairpin RNA causes interference with plant virus infection and is overcome by a virus-encoded suppressor of gene silencing. *Mol. Plant Microbe Interact.* **16**, 149–158.
15. Voinnet, O., Rivas, S., Mestre, P., and Baulcombe, D. (2003) An enhanced transient expression system in plants based on suppression of gene silencing by the p19 protein of tomato bushy stunt virus. *Plant J.* **33**, 949–956.
16. Kapila, J., de Rycke, R., van Montagu, M., and Angenon, G. (1997) An *Agrobacterium*-mediated transient gene expression system for intact leaves. *Plant Sci.* **122**, 101–108.
17. Wroblewski, T., Tomczak, A., and Michelmore, R. (2005) Optimization of *Agrobacterium*-mediated transient assays of gene expression in lettuce, tomato and *Arabidopsis*. *Plant Biotechnol. J.* **3**, 259–273.
18. Koscianska, E., Kalantidis, K., Wypijewski, K., Sadowski, J., and Tabler, M. (2005) Analysis of RNA silencing in agroinfiltrated leaves of *Nicotiana benthamiana* and *Nicotiana tabacum*. *Plant Mol. Biol.* **59**, 647–661.
19. Tang, W. and Leisner, S. (1998) Methylation of non-integrated multiple copy DNA in plants. *Biochem. Biophys. Res. Commun.* **245**, 403–406.
20. Brodersen, P., Sakvarelidze, A. L., Bruun-Rasmussen, M., Dunoyer, P., Yamamoto, Y. Y., Sieburth, L., and Voinnet, O. (2008) Widespread translational inhibition by plant miRNAs and siRNAs. *Science* **320**, 1185–1190.
21. Ruiz-Ferrer, V. and Voinnet, O. (2009) Roles of plant small RNAs in biotic stress responses. *Annu. Rev. Plant Biol.* **60**, 485–510.
22. Komari, T., Takakura, Y., Ueki, J., Kato, N., Ishida, Y., and Hiei, Y. (2006) Binary vectors and super-binary vectors. *Methods Mol. Biol.* **343**, 15–41.
23. Smith, N. A., Singh, S. P., Wang, M. B., Stoutjesdijk, P. A., Green, A. G., and Waterhouse, P. M. (2000) Total silencing by intron-spliced hairpin RNAs. *Nature* **407**, 319–320.
24. Helliwell, C. A. and Waterhouse, P. M. (2005) Constructs and methods for hairpin RNA-mediated gene silencing in plants. *Methods Enzymol.* **392**, 24–35.
25. Wise, A. A., Liu, Z., and Binns, A. N. (2006) Three methods for the introduction of foreign DNA into *Agrobacterium*. *Methods Mol. Biol.* **343**, 43–53.
26. Schöb, H., Kunz, C., and Meins, F., Jr. (1997) Silencing of transgenes introduced into leaves by agroinfiltration: a simple, rapid method for investigating sequence requirements for gene silencing. *Mol. Gen. Genet.* **256**, 581–585.
27. Bhaskar, P. B., Venkateshwaran, M., Wu, L., Ané, J. M., and Jiang, J. (2009) *Agrobacterium*-mediated transient gene expression and silencing: a rapid tool for functional gene assay in potato. *PLoS One* **4**, e5812.
28. Hofmann, T., Kalinowski, G., and Schwab, W. (2006) RNAi-induced silencing of gene expression in strawberry fruit (*Fragaria × ananassa*) by agroinfiltration: a rapid assay for gene function analysis. *Plant J.* **48**, 818–826.
29. Zottini, M., Barizza, E., Costa, A., Formentin, E., Ruberti, C., Carimi, F., and Lo Schiavo, F. (2008) Agroinfiltration of grapevine leaves for fast transient assays of gene expression and for long-term production of stable transformed cells. *Plant Cell Rep.* **27**, 845–853.
30. Kalantidis, K., Schumacher, H. T., Alexiadis, T., and Helm, J. M. (2008) RNA silencing movement in plants. *Biol. Cell* **100**, 13–26.

Chapter 8

Direct Transfer of Synthetic Double-Stranded RNA into Protoplasts of *Arabidopsis thaliana*

Ha-il Jung, Zhiyang Zhai, and Olena K. Vatamaniuk

Abstract

Double-stranded (ds) RNA interference (RNAi) is widely used as a reverse genetic approach for functional analysis of plant genes. Constitutive or transient RNAi effects in plants have been achieved via generating stable transformants expressing dsRNAs or artificial microRNAs (amiRNAs) *in planta* or by viral-induced gene silencing (VIGS). Although these tools provide outstanding resources for functional genomics, they require generation of vectors expressing dsRNAs or amiRNAs against targeted genes, transformation and propagation of transformed plants, or maintenance of multiple VIGS lines and thus impose time, labor, and space requirements. As we showed recently, these limitations can be circumvented by inducing RNAi effects in protoplasts via transfecting them with in vitro-synthesized dsRNAs. In this chapter we detail the procedure for transient gene silencing in protoplasts using synthetic dsRNAs and provide examples of approaches for subsequent functional analyses.

Key words: In vitro-synthesized dsRNA, protoplasts, protoplast viability assays, RNAi.

1. Introduction

Double-stranded RNA (dsRNA)-induced gene expression silencing [alias RNA interference (RNAi)] has been widely used as an RNA-based reverse-genetic approach in various organisms including vertebrate, invertebrate animals, and plants (1–7). The RNAi mechanism involves processing of dsRNA into short interfering RNAs (siRNAs) that guide RNase complex toward complementary motifs in single-stranded (ss) target RNAs (8). The result is the sequence-specific silencing of gene expression either at the transcriptional, mRNA stability, or translational levels (2, 9). The sequence specificity of RNAi allows silencing of individual genes as well as several genes simultaneously (2, 9).

H. Kodama, A. Komamine (eds.), *RNAi and Plant Gene Function Analysis*, Methods in Molecular Biology 744, DOI 10.1007/978-1-61779-123-9_8, © Springer Science+Business Media, LLC 2011

In plant species, constitutive RNAi effects have been achieved by transforming plants with constructs generating dsRNAs that are homologous to targeted genes, or with constructs expressing artificial microRNAs (amiRNAs) (7, 10). Transient RNAi effects have been achieved by virus-induced gene silencing (VIGS) that uses viral vectors carrying a fragment of a gene of interest for generating dsRNA (11–13). These tools, however, require generation of vectors expressing dsRNAs or amiRNAs, plant transformation and transgenic plants propagation, or maintenance of VIGS lines and thus impose time, labor, and space requirements. To circumvent these limitations, we developed a procedure for initial RNAi-based analysis of gene function using plant protoplasts and in vitro-synthesized dsRNA (14). Here we detail procedures for the design and synthesis of dsRNA, protoplast isolation, and transfection with synthetic dsRNA and provide examples of subsequent functional analyses of RNAi-silenced genes.

Advantages of RNAi in protoplasts over other RNAi-based approaches include independence of cloning of dsRNAs into plant expression or viral vectors, the circumvention of plant transformation, the methodological simplicity, the ability to knockdown multiple genes in parallel, and speediness of results. Indeed, only 20 days are sufficient for analysis. This time includes growing *Arabidopsis*, isolation of protoplasts, in vitro dsRNA synthesis, transfection of dsRNA into protoplasts, and analysis of phenotypes. The protocol has been developed for *Arabidopsis thaliana*; however, since protoplasts can be isolated from different tissues and different plants species, direct transfer of synthetic dsRNA into protoplasts can be employed as a gene silencing tool to study tissue-specific processes in a variety of species and can be adapted to a high-throughput format.

2. Materials

2.1. Preparing Plant Material for Protoplast Isolation

We typically culture plants on 1/2 concentrated Murashige and Skoog (MS) medium, pH 5.7, supplemented with 1% sucrose and 0.7% agar:

1. Murashige and Skoog (MS) basal salt mixture (Sigma). Store at 4°C.
2. Agar.
3. Petri dishes (150 mm×15 mm).
4. 70% (v/v) Ethanol.
5. Seed sterilizing bleach-Tween solution (SSBTS): 1.8% (v/v) bleach (made up by diluting household Clorox, containing 6% sodium hypochlorite), 0.1% (v/v) Tween-20 (Sigma).
6. Sterile deionized water.

2.2. Oligo Design and Generation of dsRNA Templates

1. *Arabidopsis* cDNA.
2. FastStart High Fidelity PCR System (Roche).
3. QIAquick® Gel Extraction Kit (Qiagen).
4. Agarose.
5. Ethidium bromide (Promega).
6. 10× TAE buffer: 400 mM Tris–acetate, 10 mM EDTA, pH 8.5.

2.3. Synthesis of dsRNA

1. MEGAscript High Yield T7 Transcription Kit (Ambion).
2. RNase-free DNase (Roche).
3. RNase T1 (Ambion).
4. RNeasy Mini Kit (Qiagen).
5. Custom-made oligos that include the minimal T7-RNA polymerase promoter sequence (TAATACGACTCACTATAGGG) on both 5′- and 3′-ends (15).
6. TRIzol® reagent (Invitrogen).
7. Isopropanol.
8. Chloroform.
9. AffinityScript™ QPCR cDNA Synthesis Kit (Agilent Technologies).
10. FastStart High Fidelity PCR System (Roche).
11. Ethidium bromide (Molecular Probes).
12. Gel-loading dye, 6×: 15% Ficoll 400, 66 mM EDTA, 0.1% SDS, 20 mM Tris–HCl, pH 8.0, 0.1% bromophenol blue. Gel-loading dyes are also commercially available.
13. Quick-Load® 100-bp DNA ladder (NEB).
14. 10× Tris–acetate–EDTA (TAE buffer): 400 mM Tris–acetate, 10 mM EDTA.
15. Agarose.

2.4. Isolation of Protoplasts

1. TVL solution: 300 mM sorbitol, 50 mM $CaCl_2$. Filter sterilize and store at 4°C or at –20°C.
2. Enzyme solution (freshly made and filter sterilized): 500 mM sucrose, 20 mM MES–KOH, pH 5.7, 20 mM $CaCl_2$, 40 mM KCl, 1% cellulase (Onozuka R-10; RPI Corp.), 1% macerozyme (R10; RPI Corp.).
3. W5 solution: 0.1% (w/v) glucose, 0.08% (w/v) KCl, 0.9% (w/v) NaCl, 1.84% (w/v) $CaCl_2$, 2 mM MES–KOH, pH 5.7. Filter sterilize and store at room temperature.
4. Razor blades.
5. Cheesecloth wipes.

6. Centrifuge tubes 15 mL, 50 mL.
7. Hemocytometer.
8. Pasteur pipettes.
9. Round-bottom, 2-mL centrifuge tubes.

2.5. Transfection of Protoplasts with dsRNA

1. W5 solution: see **Section 2.4**, item 3.
2. MMG solution: 4 mM MES–KOH, pH 5.7, 400 mM mannitol, 15 mM $MgCl_2$. Filter sterilize, store, and use at room temperature.
3. PEG–calcium solution: 40% PEG 4000, 200 mM mannitol, 100 mM $CaCl_2$. Filter sterilize and store at room temperature.

2.6. qRT-PCR Analysis of RNAi

1. TRIzol® reagent (Invitrogen).
2. Isopropanol.
3. Chloroform.
4. DNase I RNase-free (Roche).
5. AffinityScript™ QPCR cDNA Synthesis Kit (Agilent Technologies).
6. iQ SYBR Green Supermix (Bio-Rad).

2.7. Examples of Approaches for Functional Analysis of Knockdown Genes

1. Plant Cell Viability Assay Kit (Sigma).
2. W5 solution: see **Section 2.4**, item 3.
3. Acetonitrile (HPLC grade).
4. Phosphoric acid.

These items are optional (unless used in other parts of the protocol) and are presented here to exemplify approaches for functional analyses of RNAi protoplasts in **Section 3.8**.

3. Methods

This section details procedures for (1) isolating protoplasts from 14-day-old seedlings, (2) choosing and PCR amplification of DNA sequences that would serve as templates for in vitro dsRNA synthesis, (3) in vitro dsRNA synthesis and purification, (4) protoplast transfection, and (5) analysis of RNAi effects. Since protoplast transfection efficiency depends on the quality of isolated protoplasts, which in turn depends on the quality of starting plant material, here we also detail conditions for growing *Arabidopsis.* The latter two procedures are also available online in a video protocol (16).

3.1. Preparing Plant Material for Protoplast Isolation

1. Fifty milligrams of *Arabidopsis* wild-type seeds (ecotype Columbia) were placed into an Eppendorf microcentrifuge tube (*see* **Note 1**).
2. To sterilize seeds, 1 mL of 70% ethanol was added to the tube and mixed with seeds by vortexing. After 2 min of incubation with ethanol, seeds were spun down and the supernatant was removed.
3. One milliliter of SSBTS solution was added to seeds and sterilizing step continued for 10 more min.
4. Seeds were washed free of SSBTS five times by spinning and replacing the supernatant with sterile water. This step has to be done in a laminar flow hood.
5. Seeds were spread onto prepared 1/2 MS medium plates (50 mg of seeds per 150-mm × 15-mm plate).
6. After stratification of seeds for 24–48 h at 4°C in the dark, plates were transferred into a dedicated growth chamber and maintained at 10-h light/14-h dark cycle and at 23°C/19°C light/dark temperature regime and 75% relative humidity. Meanwhile plants were growing, dsRNA was synthesized as described below.

3.2. Oligo Design for PCR Amplification of dsRNA Templates

DNA-dependent RNA polymerases (RNAPs) that are encoded by bacteriophage T7 and its relatives (e.g., T3, SP6) are highly specific for their individual promoter sequences (17). To enable in vitro dsRNA synthesis using T7 RNA polymerase, minimal T7 RNA polymerase promoter sequences were introduced at the 5′-end of forward and reverse amplification primers followed by 17–22 gene-specific nucleotides (**Fig. 8.1a**). To increase the yield of RNA by allowing more efficient polymerase binding and transcription initiation, extra 5–6 bases were included upstream of the minimal T7 RNA polymerase promoter (**Fig. 8.1a**) (17). The flexibility in choosing extra bases also allowed achieving the desired T_m and GC content of primer pairs.

Due to the sequence specificity of RNAi-based gene expression silencing, by choosing DNA sequences that are unique or shared by related genes as dsRNA templates, it is possible to silence genes in protoplasts individually or simultaneously (14).

To exemplify the design of dsRNA templates for silencing genes individually and simultaneously, we used here *PCS1* (encoding *A. thaliana* *p*hyto*c*helatin *s*ynthase *1*) and *HXK1* and *HXK2* (encoding *A. thaliana* *he*x*o*k*inase *1*, *2*) because of the following: two *PCS* genes (*PCS1* and *PCS2*) sharing greater than 70% amino acid sequence identity are encoded by *Arabidopsis* genome (18); six members of hexokinase family, among which HXKl and HXK2 share 85% amino acid sequence, have been identified (19). To silence genes individually or simultaneously,

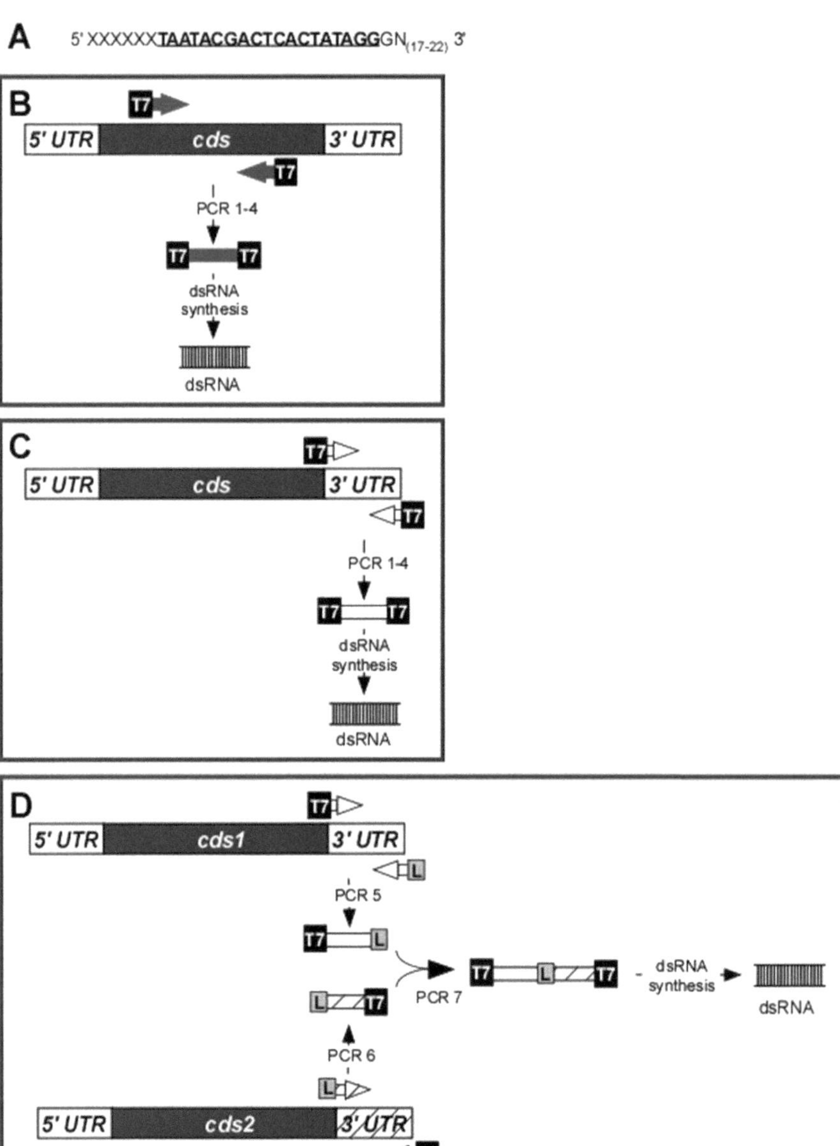

Fig. 8.1. Strategies for generating DNA templates with T7 promoter sequences by PCR. **a** Introduction of the minimal T7 RNA polymerase promoter sequence at the 5′-end of forward and reverse amplification primers. Minimal T7 RNA polymerase promoter nucleotide sequence is *underlined* and is followed by 17–22 gene-specific sequences, shown as N (17–22). Extra 5–6 bases upstream of the minimal T7 RNA polymerase promoter are shown as X. **b** Outline of the protocol for the production of dsRNA for silencing of closely related genes. 5′UTR and 3′UTR are 5′- and 3′-end untranslated regions of the gene of interest, while cds is the coding sequence. Explanation of different steps of the protocol is in the main body of the manuscript. **c** Outline of the protocol for the production of dsRNA for gene-specific silencing. 5′UTR and 3′UTR are 5′- and 3′-end untranslated regions of the gene of interest, while cds is the coding sequence. Explanation of different steps of the protocol is in the main body of the manuscript. **d** Outline of protocol for the generation of a hybrid DNA template for RNAi silencing of two genes simultaneously. In the first round of PCR (PCR reactions 5 and 6, **Table 8.1**), primer pairs (primers E1/E2 and F1/F2 in **Table 8.1**) are designed such that T7 promoter sequences (T7) at the 5′- and 3′-ends of primer pairs are followed by the gene-specific sequence corresponding to 3′UTR regions and 28 base long linker sequence (*L*, GCCTCTCTATGTACTATTGTGACGGTCC). The resulting PCR products were ligated in the second round of PCR [PCR reaction 7, primer pairs E1/F2 (**Table 8.1**)] to yield a hybrid DNA template for subsequent dsRNA synthesis.

we used the following guidelines (*see* **Note 2**): (1) to silence several closely related genes, we selected a stretch of DNA coding sequence that shares high degree of identity with related genes as RNAi targets (**Fig. 8.1b**, **Table 8.1**); (2) to silence individual genes of closely related members of a gene family, we chose 3′UTR regions as RNAi targets (**Fig. 8.1c**, **Table 8.1**); (3) to target several genes within a highly homologous multigene family, we synthesized a hybrid DNA, where 3′UTR sequences of two (or more) targeted genes were ligated by overlapping PCR (**Fig. 8.1d**, **Table 8.1**).

3.3. PCR Amplification of dsRNA Templates

Templates for dsRNA were synthesized by PCR using *Arabidopsis* cDNA as a template. However, advantage can be taken from the availability of *Arabidopsis* DNA resources listed at http://www.arabidopsis.org/portals/clones_DNA/index.jsp.

All PCR reactions should be carried out using a proofreading DNA polymerase to avoid PCR errors. We use FastStart High Fidelity PCR System (Roche) since it gives us the best yield and quality of PCR products; however, proofreading polymerases from other companies have also been working well in our hands. The PCR reactions (**Fig. 8** and **Table 8.1**) are assembled according to manufacturer's recommendations:

1. For PCR reactions from 1 to 5 in **Fig. 8** and **Table 8.1**, 2.5 μL of cDNA was used as a template in a total volume of 50 μL containing 400 nM each PCR primer, 200 μM each dNTP, and 2.5 U of DNA polymerase mix in a FastStart High Fidelity Reaction Buffer containing 1.8 mM $MgCl_2$. If primer's GC content was >50%, DMSO would be added to a final concentration of 2%.

2. PCR was carried out using MyCycler Thermal Cycler. The thermal cycling parameters were as follows: denaturation at 95°C for 2 min, followed by 30 cycles of 95°C for 30 s, 57°C for 30 s, and 72°C for 1 min.

3. PCR results were examined by 1.8% agarose gel electrophoresis to verify that a single PCR product of the expected size was generated.

4. The PCR product was isolated from gel using standard extraction procedures. We typically use QIAquick® Gel Extraction Kit (Qiagen) and elute DNA fragments in 20 μL of DNase-/RNase-free H_2O.

5. To generate hybrid RNAi targets, fusion PCR (PCR reaction 7 in **Table 8.1**, **Fig. 8.1d**) was performed by using 0.5 μL each gel-purified fragments (5) and (6) as fusion PCR templates (**Table 8.1**) in a total volume of 50 μL containing 400 nM each PCR primer, 200 μM each dNTP, and 2.5 U of DNA polymerase mix in a FastStart High Fidelity

Table 8.1
Oligonucleotide combinations for each PCR reaction and the expected size of PCR products

Target	Template	Forward oligo name, sequence	Reverse oligo name, sequence	PCR reaction	PCR product, length (bp)
PCS1+3′ UTR	$cDNA_{PCS1}$	A1 GCGCCTAATACGACTCACT ATAGGGAGAGAACCTCTGG AAGTAGTGAAGGAA	A2 GCGCCTAATACGACTCACT ATAGGGAGACTGTGAACTT ACAAGACGAGGAAC	1	547
HXK1/HXK2	CDS_{HXK1}	B1 AAGCTTTAATACGACTCACT ATAGGGCTGCTTCCAAAAT CAGGAGAAA	B2 GGATCCTAATACGACTCACT ATAGGGGTAATGCTCAAACAA TCCACCA	2	524
HXK1	$3'UTR_{HXK1}$	C1 GCGCCTAATACGACTCACT ATAGGGAGATCGCTATCAG AAAACGCCTAA	C2 GCGCCTAATACGACTCACT ATAGGGAGACTCCAGTGAA GTGAGCTTTGA	3	281
HXK2	$3'UTR_{HXK2}$	D1 GCGCCTAATACGACTCACT ATAGGGAGATTGTTTTGCCG TTAGGGT	D2 GCGCCTAATACGACTCACT ATAGGGAGAACAACATTGG CTGGTCGTTT	4	207
HXK1+linker	$3'UTR_{HXK1}$	E1 GCGCCTAATACGACTCACT ATAGGGAGATCGCTATCAG AAAACGCCTAA	E2 GGACCGTCACAATAGTACA TAGAGAGGCCTCCAGTGAA GTGAGCTTTGA	5	309
HXK2+linker	$3'UTR_{HXK2}$	F1 GCCTCTCTATGTACTATTGT GACGGTCCTTGTTTTGCCGT TAGGGTTT	F2 GCGCCTAATACGACTCACT ATAGGGAGAACAACATTGG CTGGTCGTTT	6	235
HXK1/HXK2	Fragments from PCR 5 and 6	E1 GCGCCTAATACGACTCACT ATAGGGAGATCGCTATCAG AAAACGCCTAA	F2 GCGCCTAATACGACTCACT ATAGGGAGAACAACATTGG CTGGTCGTTT	7	516

Reaction Buffer containing 1.8 mM $MgCl_2$. PCR was carried out using MyCycler Thermal Cycler. The thermal cycling parameters were as follows: denaturation at 95°C for 3 min, followed by 30 cycles of 95°C for 30 s, 52°C for 30 s, and 72°C for 30 s. PCR results were examined by 1.8% agarose gel electrophoresis, PCR product of the expected size was gel purified using the QIAquick® Gel purification kit, and the DNA fragment was eluted in 20 μL of DNase-/RNase-free H_2O.

3.4. Synthesis of dsRNA

dsRNA was synthesized in vitro using the MEGAscript T7 kit:

1. Eight microliters of the DNA template (the purified PCR product from **Section 3.3**) was added into a sterile, nuclease-free tube containing 2 μL of 10× transcription buffer, 2 μL of rNTPs (7.5 mM each), and 2 μL of RNA polymerase mix (5U/μL) (MEGAscript T7 kit).
2. Volume of the mixture was adjusted to 20 μL with nucleases-free water.
3. Components were mixed by vortexing and the reaction mixture was collected at the bottom of the tube by spinning down for several seconds in a Galaxy MiniStar minicentrifuge (VWR International).
4. RNA transcription reaction proceeded for 16 h (overnight) at 37°C. The transcribed sense and antisense RNA strands had self-annealed.
5. To remove DNA template and ssRNA (*see* **Note 3**), 1 μL of RNase-free DNase I and 1 μL of RNase T1 were added to the reaction mixture.
6. The reaction mixture was incubated at 37°C for 30 min.
7. dsRNA was purified using RNeasy kit (Qiagen) according to manufacturer's recommendations. Alternatively, dsRNA can be purified by precipitating in sodium acetate, pH 5.2, and ethanol or isopropanol.
8. To determine dsRNA concentration, dsRNA was diluted 50- to 100-fold and the absorbance was read at a wavelength of 260 nm (*see* **Note 4**).
9. To verify concentration and the relative size, dsRNA was analyzed using native agarose gel electrophoresis (*see* **Note 4**). To do so, 500 ng of dsRNA in 1× loading buffer was subjected to electrophoresis on the agarose gel containing 1.2% (w/v) agarose, 40 mM Tris–acetate, 1 mM EDTA (1× TAE buffer), and 0.5 μg/mL ethidium bromide. The representative results produced are shown in **Fig. 8.2.**).

3.5. Isolation of Protoplasts

1. Fourteen-day-old *Arabidopsis* seedlings were collected from Petri dishes (*see* **Note 5**).

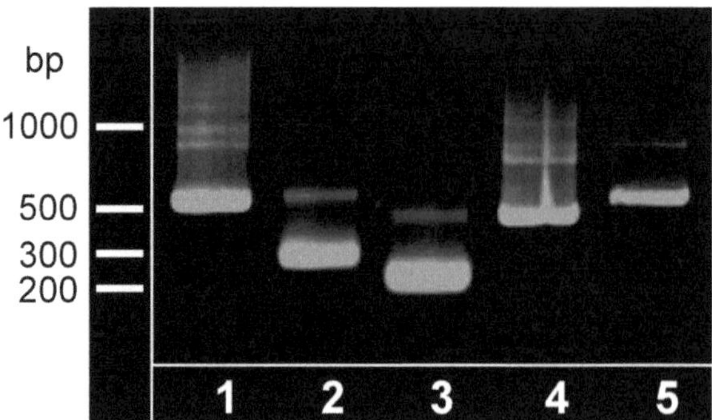

Fig. 8.2. Native agarose gel analysis of different sized dsRNA molecules generated using the MEGAscript T7 kit. Approximately 500 ng of each dsRNA was separated on a 1.8% agarose, 0.5 μg/mL ethidium bromide/1× TAE gel. Lane designations: *lane 1*, *HXK1/HXK2* (PCR reaction 2, **Table 8.1**); *lane 2*, *HXK1*+linker (PCR reaction 5, **Table 8.1**); *lane 3*, *HXK2*+linker (PCR reaction 6, **Table 8.1**); *lane 4*, *HXK1/HXK2* (PCR reaction 7, **Table 8.1**); and *lane 5*, *PCS1+3′ UTR* (PCR reaction 1, **Table 8.1**). Also show the migration of 100-bp DNA ladder.

2. Two grams of plant material was sliced with a razor blade in 15 mL of filter-sterilized TVL Solution (*see* **Note 6**).
3. Chopped tissues were transferred into a 200-mL beaker containing 20 mL of filter-sterilized enzyme solution.
4. Plant material was mixed with enzyme solution by gently swirling the beaker; the beaker was covered with parafilm and aluminum foil and cell wall digestion proceeded in the dark at room temperature for 16–18 h with gentle shaking at 35 rpm.
5. The released protoplasts were collected into a 50-mL Falcon tube by sieving through eight layers of the cheesecloth, pre-wet in W5 Solution. Protoplasts were sieved from the cheesecloth one more time into the same 50-mL Falcon tube by washing the cloth with 10 mL of W5 solution.
6. Protoplasts were overlaid with 5 mL of W5 solution (*see* **Note 7**) and centrifuged for 7 min at 100×*g*.
7. Ten milliliters of protoplasts was collected at the interface of enzyme and W5 solutions (**Fig. 8.3a**) and transferred to a new 50-mL Falcon tube.
8. To wash protoplasts free from enzyme solution, 15 mL of W5 solution was added to a tube with protoplast, mixed gently, and centrifuged for 5 min at 60×*g*. After removing the supernatant, protoplasts were washed one more time by resuspending in 10 mL of W5 solution.
9. After collecting protoplasts by centrifugation for 5 min at 60×*g*, the supernatant was removed and final protoplast pellet was reconstituted in 1–3 mL of W5 solution.

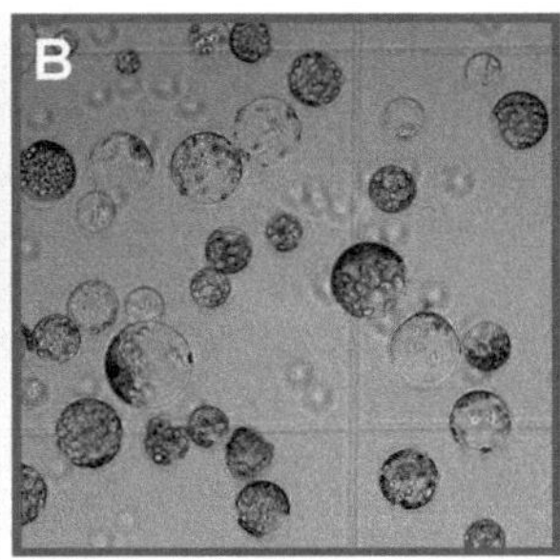

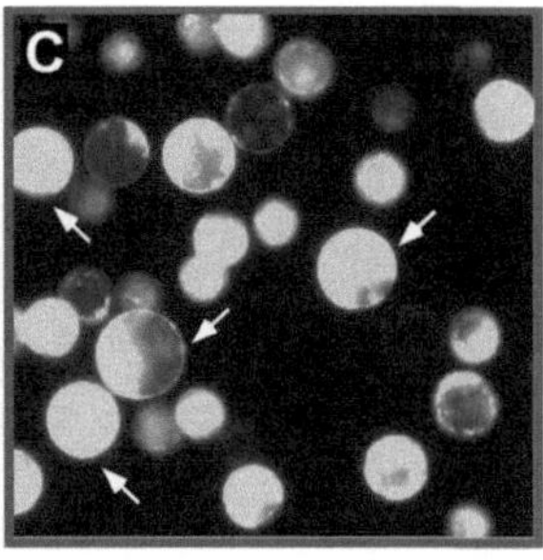

Fig. 8.3. Isolation of protoplasts from *Arabidopsis* seedlings. **a** Protoplasts were purified by sucrose density gradient centrifugation and collected at the interface of enzyme solution and W5 buffer (*white arrows*). **b** Bright-field microscopy of protoplasts. **c** Epifluorescent imaging of protoplast viability after staining with fluorescein diacetate. Fluorescein diacetate is a lipophilic, membrane-permeable, and non-fluorescent dye that freely diffuses into protoplasts. In viable protoplasts, it is hydrolyzed to a highly polar, membrane-impermeable fluorescent compound that is retained within viable protoplasts, producing an intense green fluorescence within the cytoplasm (*white arrows*). Microphotographs were collected using a cooled CCD camera interfaced with the Zeiss Axioskop 2 plus microscope.

10. Protoplast yield was evaluated by cell counting with a hemocytometer. Protoplast viability was assessed using fluorescein diacetate from Plant Cell Viability Assay Kit (Sigma), according to manufacturer's recommendations. This procedure typically yields up to 5×10^6 viable protoplasts from 1 g of fresh seedlings. An example of results produced is shown in **Fig. 8.3b, c**.

3.6. Transfection of Protoplasts with dsRNA

1. Protoplasts, suspended in W5 medium (from **Section 3.5**, item 8), were placed on ice for 30 min; protoplasts sediment at the bottom of the tube.
2. W5 medium was removed by pipetting out the majority of the supernatant and replaced with 1–2 mL aliquot of MMG solution (*see* **Note 8**).
3. A 100 μL aliquot of protoplasts was transferred into a 2-mL round-bottom microcentrifuge tube before 10 μL of dsRNA (1 μg/μL) was added into the tube. Protoplasts and dsRNA were mixed by gently tapping the bottom of the tube.
4. The transfection was started by adding 110 μL of PEG–calcium transfecting solution and mixing the components of the reaction by gently tapping the tube.
5. The transfection mixture was incubated for 7 min at room temperature.
6. The reaction was terminated by diluting the transfection mixture with 600 μL of W5 solution and mixing by slowly inverting the tube.

7. The transfected protoplasts were washed free of remaining dsRNA and PEG–calcium solution. To do so, protoplasts were pelleted at 100×*g* for 2 min at room temperature. The supernatant (770 μL) was removed; the remaining 50 μL of protoplasts were used for subsequent manipulations.
8. Nine hundred and fifty microliters of W5 solution was added to the protoplast suspension to reach a total volume of 1,000 μL.
9. The transfected protoplasts were incubated at room temperature in the dark from 24 to 96 h before analyzing RNAi effects (*see* **Note 9**).

3.7. qRT-PCR Analysis of RNAi

At the onset of functional analysis, it is important to evaluate the extent of the depletion of the RNAi-targeted transcript. It is also recommended to test the specificity of RNAi effects by analyzing the transcript level of related genes. We used quantitative real-time (qRT) PCR for measuring the transcript levels. To ensure reproducible and accurate measurements, RNA sample preparation, reverse transcription, and quantitative PCR were carried out using recently proposed guidelines (21, 22). We showed previously that the depletion of RNAi-targeted transcript is observed 24 h of transfection and that RNAi effects last for at least 96 h (14) (**Fig. 8.4**). Therefore, we suggest evaluating the abundance of RNAi-targeted transcript after 24 h of dsRNA transfection.

qRT-PCR analysis of the abundance of knockdown genes consists of the following steps: RNA isolation from RNAi protoplasts, cDNA synthesis, and quantitative PCR analysis (or semi-quantitative RT-PCR (14)):

1. Transfected protoplasts (from **Section 3.6**, item 9) were concentrated by centrifugation at 100×*g* for 2 min.
2. Supernatant (900 μL) was discarded and 1 mL of TRIzol® reagent (Invitrogen) was added into remaining 100 μL of the protoplast mixture (containing at least 10^5 protoplasts). RNA was isolated using manufacturer's recommendations. RNA pellet was reconstituted in 10 μL of nuclease-free water and stored RNA at –80°C. Because of the presence of phenol in TRIzol®, and subsequent use of chloroform, RNA isolation should be performed in the fume hood.
3. One microgram of total RNA template, 10 μL of 2× first-strand master mix (AffinityScript™ QPCR cDNA Synthesis Kit), and 1 μL of amplification-grade, RNase-free DNase I (10 U/μL) were added on ice into a sterile, nuclease-free tube and the volume was adjusted to 16 μL with nuclease-free water.
4. The DNase treatment reaction was performed at 37°C for 20 min (*see* **Note 10**).

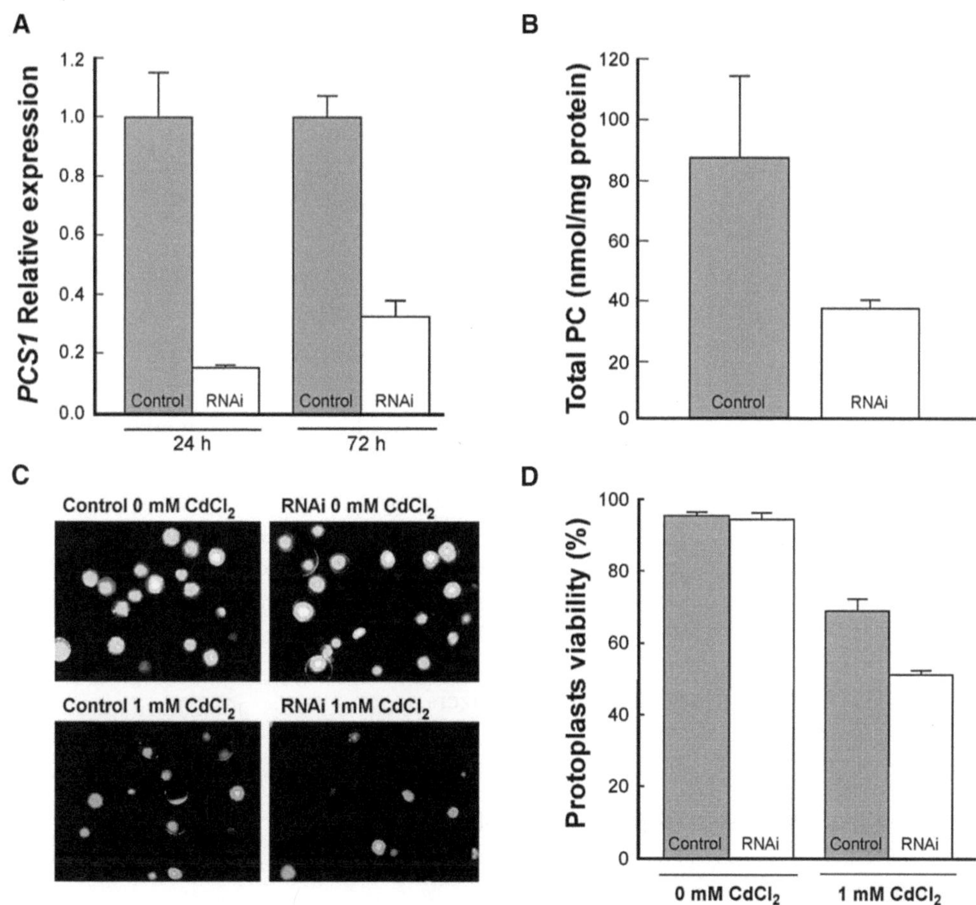

Fig. 8.4. Functional analysis of *PCS1* by RNAi in protoplasts. **a** Quantitative real-time PCR analysis (qRT-PCR) of the abundance of *PCS1* in protoplasts transfected with $dsRNA_{PCS1}$ (RNAi) or water (control). Total RNA was extracted from protoplasts after 24 or 72 h of transfection. First-strand cDNA synthesis was carried out using AffinityScript^{TM} QPCR cDNA Synthesis Kit. PCR reactions were performed in a CFX96 real-time PCR detection system using iQ SYBR Green Supermix to monitor dsDNA synthesis. At least three biological replicates, with technical duplicates for each biological replicate, were performed. The relative *PCS1* transcript level was determined for each sample after normalizing to the level of *ACTIN 2* cDNA. Error bars represent SE. **b** Reverse-phase HPLC analysis of total phytochelatin (PC) content in protoplasts transfected with water (control) or with $dsRNA_{PCS1}$ (RNAi). Cadmium was added as $CdCl_2$ (1 mM) to the protoplast culture after 24 h of transfection. Protoplasts were incubated in the dark at room temperature for 48 h and then collected by centrifugation. Protoplasts were lysed by vortexing in 200 μL of the lysis buffer containing 50 mM Tris–HCl, pH 8.0, 1 mM 2-mercaptoethanol, 1 mM EGTA, and protease inhibitors. Protein concentration of samples was estimated in 20 μL aliquots using dye-binding method (28). PC content was analyzed by RP-HPLC as described previously (27). Results represent an average of three biological replicates. Error bars represent SD. **c**, **d** Viability of mock-transfected (control) and $dsRNA_{PCS1}$ (RNAi)-transfected protoplasts cultured in W5 medium (0 mM $CdCl_2$) or in W5 supplemented with 1 mM $CdCl_2$ (1 mM $CdCl_2$). Cadmium was added after 24 h of transfection and protoplasts were cultured for 48 h in the dark; then protoplasts were collected by centrifugation and stained with fluorescein diacetate. Viable protoplasts exhibiting intense fluorescence were counted with hemocytometer. **c** Epifluorescent imaging of protoplast viability. Images were captured using Zeiss Axioskop 2 plus microscope equipped with a GFP filter set. Examples of viable protoplasts are indicated by *white arrows*. **d** Evaluation of protoplast viability by counting fluorescent cell using a hemocytometer. The figure represents the mean values from counting viable cells in five biological and three technical replicates for each biological replicate. Error bars represent SE.

5. DNase I was heat inactivated by incubating the reaction at 65°C for 15 min.
6. The reaction was cooled on ice for 2 min before 3 μL of oligo $(dT)_{18}$ primer (0.1 μg/μL) and 1 μL of AffinityScript RT/RNase block enzyme mixture (AffinityScript™ QPCR cDNA Synthesis Kit) were added to the tube. The final reaction mixture was 20 μL.
7. To allow primer annealing, the reaction was incubated at 25°C for 5 min.
8. First-strand cDNA synthesis was performed at 42°C for 30 min.
9. The reaction was terminated by inactivating enzyme at 95°C for 5 min.
10. One microliter of the first-strand cDNA was used as a template for quantitative PCR in a total volume of 15 μL containing 400 nM of each PCR primer (**Table 8.2**), 50 mM KCl, 20 mM Tris–HCI, pH 8.4, 0.2 mM each dNTP and 1.25 U of iTaq DNA polymerase in iQ SYBR Green Supermix, containing 3 mM $MgCl_2$, SYBR Green I, 20 nM fluorescein, and stabilizers.

Table 8.2
Oligonucleotide sequences used for qRT-PCR

Size of PCR product (bp)	Targeted gene	Direction	Oligo (5′–3′)
153	*PCS1*	Forward	CCCAATTTCGAATCTCACACACCG
		Reverse	GAGATCGCCGATATAAACTCGCCA
141	*ACT2*	Forward	GACCTTTAACTCTCCCGCTA
		Reverse	GGAAGAGAGAAACCCTCGTA

11. PCR was carried out using CFX96 real-time PCR system (Bio-Rad). The thermal cycling parameters were as follows: denaturation at 95°C for 3 min, followed by 39 cycles of 95°C for 10 s, 60°C for 30 s. Amplicon dissociation curves, i.e., melting curves, were recorded after cycle 39 by heating from 60 to 95°C with 0.5°C increments and an average ramp speed of 3.3°C s^{-1}.
12. Data were analyzed using CFX Manager software, version 1.5 (Bio-Rad). *Arabidopsis ACTIN 2* was used as a reference for normalizing gene expression data. The representative results are shown in **Fig. 8.4a**.

3.8. Examples of Approaches for Functional Analysis of Knockdown Genes

Since protoplasts are non-growing cells, effective RNAi-triggered gene silencing depends not only on transcripts depletion but also on turnover rates of corresponding polypeptides. Using relatively abundant gene, *ECS1* (alias *cad2*), encoding γ-glutamylcysteine synthase (γ-ECS) of *Arabidopsis* as an RNAi target, we showed that its polypeptide level was diminished after 72 h of dsRNA transfection (14). Therefore, we suggest performing functional analysis of RNAi protoplasts after 72–96 h of transfection.

We have discussed previously different applications as well as limitations of RNAi in protoplasts (see **Note 11** and ref. 14). To exemplify approaches for functional analysis of knockdown genes in protoplasts, we used *PCS1* of *Arabidopsis*, encoding *p*hyto*c*helatin *s*ynthase *1* with well-established role in heavy metal detoxification (18, 23). In the presence of heavy metals, *PCS1* catalyzes the formation of small, cysteine-rich peptides, phytochelatins (PCs) from a tripeptide glutathione (GSH). PCs chelate heavy metals and facilitate vacuolar heavy metal sequestration and long-distance transport (24–26). *PCS1*-deficient *cad1-3 Arabidopsis* knockout line is heavy metal hypersensitive and lacks PC (23). Therefore, we expected that heavy metal-treated [e.g., cadmium (Cd)-treated), $dsRNA_{PCS1}$-transfected protoplasts will accumulate less PC compared to Cd-treated mock-transfected protoplasts. We also expected that Cd tolerance of RNAi protoplasts will be reduced.

We knockdown *PCS1* in *Arabidopsis* protoplasts using the above-described procedure. After verifying the depletion of *PCS1* transcript by qRT-PCR (**Fig. 8.4a**), we performed the following assays: first, we analyzed the PC content in Cd-treated RNAi and control protoplasts using reverse-phase (RP) HPLC (27); second, we evaluated the viability of Cd-treated control and RNAi protoplasts using staining with fluorescein diacetate (Plant Cell Viability Assay Kit).

Reverse-phase (RP) HPLC analysis showed that PC content in $CdCl_2$-cultured RNAi protoplasts was 2.4-fold lower than that of mock-transfected protoplasts cultured in the same conditions (**Fig. 8.4b**). Since, PCS1-catalyzed PC formation from GSH depends on the presence of Cd, as expected, we did not detect PCs in RNAi or control protoplasts cultured without this heavy metal (not shown). The fold decrease in total PC content in RNAi protoplasts after culturing for 48 h with 1 mM $CdCl_2$ (72 h after $dsRNA_{PCS1}$ transfection) was consistent with 2.5–3-fold reduction of *PCS1* transcript level after 72 h of transfection (**Fig. 8.4a, b**). Finally, as expected, silencing of *PCS1* in protoplasts increased their Cd sensitivity (**Fig. 8.4c, d**). These data demonstrate that RNAi can be used as a versatile approach to study gene function in protoplasts.

To conclude, RNAi in protoplasts described here is a rapid, affordable, and potent reverse-genetic approach that is complementary to existing genetic tools as it allows expediting functional analysis of plant genes and selecting candidates for subsequent studies *in planta*.

4. Notes

1. Fifty milligrams of seeds was spread onto one 150-mm ×15-mm plate; after germination and culturing for 14 days, approximately 4 g of healthy seedlings was obtained.
2. We observed that a synthetic dsRNA of 400 bp or larger (up to 1 kb) provides the strongest silencing effects (unpublished observations). The minimum size of dsRNA recommended for RNAi is ~200 bp.
3. If accurate measurement of dsRNA concentration is required, a DNA template and ssRNA should be removed from the reaction mixture. Since RNase T1 cleaves ssRNA with high specificity at guanylyl residues (20), RNase T1 treatment will remove any remaining ssRNA leaving dsRNA intact.
4. We showed previously that the transcript depletion efficiency in protoplasts depends on the dose of the transfected dsRNA (14). Therefore, it is important and necessary to quantify dsRNA prior subsequent steps. We assess dsRNA concentration using the traditional UV spectroscopy method (one A260 unit equals ~40 μg/mL of dsRNA). We also visualize dsRNA using native agarose gel electrophoresis. The concentration of agarose depends on the size of the transcript (1–2.0% agarose for transcripts from 1,000 to 200 bp). Including DNA ladder on the gel can help to determine the size and quantity of dsRNA. Note that dsRNA might migrate slower on a native agarose gel. In addition, incorporation of ethidium bromide into the gel could also affect dsRNA migration properties. Therefore, you may choose not to add ethidium bromide into the gel but instead stain the gel after the electrophoresis. If accurate determination of dsRNA size is needed, estimate the transcript size by RNA denaturing gel.
5. As we emphasized in a video protocol for isolating protoplasts (16), to ensure the high yield of intact protoplast, it is very important to start with healthy plants. Use filter-

sterilized solutions for isolating protoplasts. Remember that protoplasts are fragile. Therefore, when handling protoplasts, do not mix, pipet, or vortex them vigorously since it will break them. Instead, mix protoplasts by slowly rotating or taping the centrifuge tube.

6. We chop plant material in a lid of the Petri dish used for culturing seedlings. The remaining plant material can be used for RNA isolation and cDNA synthesis.
7. When overlaying protoplasts with W5 solution, care should be taken not to disturb the sugar gradient.
8. MMG solution should be of room temperature. The volume of MMG solution may vary depending on the concentration of isolated protoplasts. For successful transformation, it is important that the final density is 1×10^6 protoplasts/mL.
9. We noticed that placing tubes with transfected protoplasts on a side instead of keeping vertically significantly improves their viability after incubation for prolonged time after transfection.
10. In order to eliminate artifacts arising from genomic DNA contamination in RNA samples, it is important to eliminate any contaminating genomic DNA prior to cDNA synthesis and analysis of the abundance of the knockdown gene transcript.
11. As we discussed in (14), limitations of using RNAi in protoplasts for functional analysis of genes include potential artifacts that might emerge due to the removal of plant cell walls, disruption of cell-to-cell communications, and the artificial protoplast culture conditions. Clearly, transient RNAi effects in protoplasts have to be verified and complemented by studies *in planta*. Nevertheless, transient RNAi in protoplasts, achieved by transfecting synthetic dsRNAs, offers a rapid, potent, methodologically simple, and affordable reverse-genetic approach that is complementary to existing genetic tools. This approach expedites functional analysis of genes, providing an initial screening step for selecting candidate genes for subsequent studies *in planta*. As exemplified here and in ref. 14 this procedure is particularly applicable for analysis of genes, which silencing could lead to conditional lethality, for biochemical studies aiming to decipher the components of metabolic pathways, and for establishing the metabolic networks. Finally, adapting RNAi in protoplasts to a high-throughput, genome-wide phenotyping technology will greatly facilitate assigning functions to unknown genes not only in *Arabidopsis* but also in other plant species.

References

1. Allen, R. S., Millgate, A. G., Chitty, J. A., Thisleton, J., Miller, J. A. C., et al. (2004) RNAi-mediated replacement of morphine with the nonnarcotic alkaloid reticuline in opium poppy. *Nat. Biotech.* **22**, 1559–1566.
2. Baulcombe, D. (2004) RNA silencing in plants. *Nature* **431**, 356–363.
3. Fire, A., Xu, S., Montgomery, M. K., Kostas, S. A., Driver, S. E., et al. (1998) Potent and specific genetic interference by double-stranded RNA in *Caenorhabditis elegans. Nature* **391**, 806–811.
4. Kennerdell, J. R. and Carthew, R. W. (1998) Use of dsRNA-mediated genetic interference to demonstrate that frizzled and frizzled 2 act in the wingless pathway. *Cell* **95**, 1017–1026.
5. Smith, N. A., Singh, S. P., Wang, M.-B., Stoutjesdijk, P. A., Green, A. G., et al. (2000) Gene expression: total silencing by intron-spliced hairpin RNAs. *Nature* **407**, 319–320.
6. Vidali, L., Augustine, R. C., Kleinman, K. P., and Bezanilla, M. (2007) Profilin is essential for tip growth in the moss *Physcomitrella patens. Plant Cell* **19**, 3705–3722.
7. Waterhouse, P. M. and Helliwell, C. A. (2003) Exploring plant genomes by RNA-induced gene silencing. *Nat. Rev. Genet.* **4**, 29–38.
8. Zamore, P. D. (2001) RNA interference: listening to the sound of silence. *Nat. Struct. Biol.* **8**, 746–750.
9. Brodersen, P., Sakvarelidze-Achard, L., Bruun-Rasmussen, M., Dunoyer, P., Yamamoto, Y. Y., et al. (2008) Widespread translational inhibition by plant miRNAs and siRNAs. *Science* **320**, 1185–1190.
10. Schwab, R., Ossowski, S., Warthmann, N., and Weigel, D. (2010) Directed gene silencing with artificial microRNAs. *Methods Mol. Biol.* **592**, 71–88.
11. Burch-Smith, T. M., Schiff, M., Liu, Y., and Dinesh-Kumar, S. P. (2006) Efficient virus-induced gene silencing in *Arabidopsis. Plant Physiol.* **142**, 21–27.
12. Dinesh-Kumar, S. P., Anandalakshmi, R., Marathe, R., Schiff, M., and Liu, Y. (2003) Virus-induced gene silencing. *Methods Mol. Biol.* **236**, 287–294.
13. Lu, R., Martin-Hernandez, A. M., Peart, J. R., Malcuit, I., and Baulcombe, D. C. (2003) Virus-induced gene silencing in plants. *Methods* **30**, 296–303.
14. Zhai, Z., Sooksa-nguan, T., and Vatamaniuk, O. K. (2009) Establishing RNA interference as a reverse-genetic approach for gene functional analysis in protoplasts. *Plant Physiol.* **149**, 642–652.
15. Sastry, S. S. and Ross, B. M. (1997) Nuclease activity of T7 RNA polymerase and the heterogeneity of transcription elongation complexes. *J. Biol. Chem.* **272**, 8644–8652.
16. Zhai, Z., Jung, H. I., and Vatamaniuk, O. K. (2009) Isolation of protoplasts from tissues of 14-day-old seedlings of *Arabidopsis thaliana. J. Vis. Exp.* doi: 10.3791/1149.
17. Rong, M., He, B., McAllister, W. T., and Durbin, R. K. (1998) Promoter specificity determinants of T7 RNA polymerase. *Proc. Natl. Acad. Sci. USA* **95**, 515–519.
18. Vatamaniuk, O. K., Mari, S., Lu, Y. P., and Rea, P. A. (1999) *AtPCS1*, a phytochelatin synthase from Arabidopsis: isolation and in vitro reconstitution. *Proc. Natl. Acad. Sci. USA* **96**, 7110–7115.
19. Karve, A., Rauh, B. L., Xia, X., Kandasamy, M., Meagher, R. B., et al. (2008) Expression and evolutionary features of the hexokinase gene family in *Arabidopsis. Planta* **228**, 411–425.
20. Heinemann, U. and Saenger, W. (1983) Crystallographic study of mechanism of ribonuclease T1-catalysed specific RNA hydrolysis. *J. Biomol. Struct. Dyn.* **1**, 523–538.
21. Remans, T., Smeets, K., Opdenakker, K., Mathijsen, D., Vangronsveld, J., et al. (2008) Normalisation of real-time RT-PCR gene expression measurements in *Arabidopsis thaliana* exposed to increased metal concentrations. *Planta* **227**, 1343–1349.
22. Udvardi, M. K., Czechowski, T., and Scheible, W.-R. (2008) Eleven golden rules of quantitative RT-PCR. *Plant Cell* **20**, 1736–1737.
23. Howden, R., Goldsbrough, P. B., Andersen, C. R., and Cobbett, C. S. (1995) Cadmium-sensitive, *cad1* mutants of *Arabidopsis thaliana* are phytochelatin deficient. *Plant Physiol.* **107**, 1059–1066.
24. Salt, D. E. and Rauser, W. E. (1995) MgATP-dependent transport of phytochelatins across the tonoplast of oat roots. *Plant Physiol.* **107**, 1293–1301.
25. Chen, A., Komives, E. A., and Schroeder, J. I. (2006) An improved grafting technique for mature *Arabidopsis* plants demonstrates long-distance shoot-to-root transport of phytochelatins in Arabidopsis. *Plant Physiol.* **141**, 108–120.
26. Mendoza-Cozatl, D. G., Butko, E., Springer, F., Torpey, J. W., Komives, E. A., et al.

(2008) Identification of high levels of phytochelatins, glutathione and cadmium in the phloem sap of *Brassica napus*. A role for thiol-peptides in the long-distance transport of cadmium and the effect of cadmium on iron translocation. *Plant J.* **54**, 249–259.

27. Vatamaniuk, O. K., Mari, S., Lu, Y. P., and Rea, P. A. (2000) Mechanism of heavy metal ion activation of phytochelatin (PC) synthase: blocked thiols are sufficient for PC synthase-catalyzed transpeptidation of glutathione and related thiol peptides. *J. Biol. Chem.* **275**, 31451–31459.
28. Bradford, M. M. (1976) A rapid and sensitive method for the quantitation of microgram quantities of protein utilizing the principle of protein-dye binding. *Anal. Biochem.* **72**, 248–254.

Chapter 9

Detection of Long and Short Double-Stranded RNAs

Toshiyuki Fukuhara, Syunichi Urayama, Ryo Okada, Eri Kiyota, and Hiromitsu Moriyama

Abstract

In RNA interference (RNAi), long double-stranded RNAs (dsRNAs) of more than 100 nucleotides (nt) are diced into short dsRNAs (small interfering RNAs, siRNAs) of about 21–24 nt, the guide strand of which is incorporated into the RNA-induced silencing complex (RISC) that slices a specific mRNA. Consequently viral dsRNAs are known as potent inducers for RNAi, which probably originated from a defense mechanism against nucleic acid parasites. Therefore detection of long and short dsRNAs must be crucial techniques for RNAi or virus research. The methods for simple and sensitive detection of short dsRNAs (siRNAs) by northern hybridization, isolation of long dsRNAs by CF-11 cellulose chromatography, and detection of long dsRNAs by agarose gel electrophoresis and northern hybridization are described here.

Key words: Agarose gel electrophoresis, CF-11 cellulose chromatography, dsRNA, northern hybridization, plasmid-like dsRNA, RNAi, siRNA.

1. Introduction

Before the discovery of RNA interference (RNAi) from nematode in 1998 (1), an interesting phenomenon called co-suppression or post-transcriptional gene silencing (PTGS) was discovered from plants in 1990 (2, 3). PTGS (co-suppression) is considered to be a similar phenomenon as RNAi, because a trigger molecule of both animal RNAi and plant PTGS is a long double-stranded RNA (dsRNA) of more than 100 nucleotides (nt). These long dsRNAs are diced into short dsRNAs (small interfering RNAs, siRNAs) of about 21–24 nt by an RNase III enzyme, Dicer, and then, Slicer (Argonaute) with a guide strand of short dsRNA (siRNA), which is termed the RNA-induced silencing complex (RISC), cleaves an mRNA in a sequence-specific manner. RNAi/PTGS

H. Kodama, A. Komamine (eds.), *RNAi and Plant Gene Function Analysis*, Methods in Molecular Biology 744, DOI 10.1007/978-1-61779-123-9_9,

probably originated from a defense mechanism against nucleic acid parasites, because a long dsRNA is a signal molecule for virus infection (4). Even all single-stranded RNA (ssRNA) viruses must go through replication intermediate dsRNAs during their replication. Viral dsRNAs are known as potent inducers for RNAi, and the accumulation of short dsRNAs (siRNAs) derived from viral dsRNAs is observed during virus infection (5). Therefore, detection of long dsRNAs and short dsRNAs (siRNAs) is one of the most crucial techniques for studying RNAi/PTGS and virus infection.

In the 1980s (before the discovery of animal RNAi and plant PTGS), it had been reported that linear dsRNAs of about 1.0–20 kbp in length were often found from various plants (6). Because these host plants exhibit neither symptoms nor abnormal phenotypes, these dsRNAs have been called plasmid-like dsRNAs or endogenous dsRNAs. These plasmid-like dsRNAs are maintained stably in host plants and inherited efficiently to the next generation of host plants via pollen or ova. The reason why these plasmid-like dsRNAs exist stably in host plants with an RNAi system remains unknown. To isolate these plasmid-like or viral dsRNAs, CF-11 cellulose column chromatography had been developed about 30 years ago (7, 8). It is still one of the most useful procedures for isolating long dsRNAs from total nucleic acids. Here we report the methods for simple and sensitive detection of siRNAs by northern hybridization, isolation of long dsRNAs by CF-11 cellulose chromatography, and detection of long dsRNAs by agarose gel electrophoresis (AGE) and northern hybridization.

2. Materials

2.1. Isolation of Total RNA from Plant Tissues

1. Absolute ethanol.
2. 70% ethanol.
3. 3 M sodium acetate, pH 5.2.
4. 5 mg/mL ethidium bromide (EtBr) stock solution (dissolved in water).
5. Agarose (electrophoresis grade).
6. Trizol reagent (Invitrogen).
7. 50X TAE (Tris–acetate–EDTA) buffer: 2 M Tris–acetate, pH 8.0, and 50 mM EDTA.
8. RNA sample buffer: 1.6 mL of formaldehyde (37% stock solution), 5.0 mL of formamide, 0.5 mL of 20X MOPS (*see* **Section 2.2**, Step 5), and 1.6 mL of 50% glycerol containing 0.1 mg/mL bromophenol blue (BPB) and 1 mM EDTA.

9. DEPC-treated water: add 0.1% diethylpyrocarbonate (DEPC) to water and stir the DEPC solution gently for more than 1 h in a draft chamber, because DEPC is toxic and volatile. Autoclave it for 40 min to remove (decompose) DEPC from water.
10. High-speed refrigerated centrifuge (max speed about $20{,}000\times g$ (15,000 rpm)).
11. Mini submerged and horizontal agarose gel electrophoresis chamber, gel size 50 mm (W) × 60 mm (L) (e.g., Advance Co., Ltd., Tokyo, Japan, Mupid-2plus).
12. Centrifugal concentrator with a vacuum pump.
13. UV spectrophotometer.
14. UV transilluminator.

2.2. Denaturing PAGE for Short dsRNAs (siRNAs)

1. Urea.
2. Thirty percent acrylamide/bis solution (acrylamide:bisacrylamide = 29:1).
3. *N,N,N,N′*-tetramethylethylenediamine (TEMED).
4. Ammonium persulfate: prepare 10% solution in water and store for about 1 month at 4°C.
5. 20X MOPS (3-morpholinopropanesulfonic acid) buffer: 0.4 M MOPS, pH 7.0 (adjusted with NaOH), 0.1 M sodium acetate, and 20 mM EDTA.
6. Formamide RNA solution: 1.0 mg/mL xylene cyanol (XC), 1.0 mg/mL BPB, and 20 mM EDTA in 96% formamide.
7. Mini-slab gel electrophoresis chamber (e.g., ATTO, Tokyo, Japan).
8. Power supply.

2.3. Transfer and Cross-Linking of siRNAs to a Nylon Membrane

1. Cross-linking solution: immediately prior to use, prepare 0.16 M l-ethyl-3-(3-dimethylaminopropyl)carbodiimide (EDC, Sigma) in 0.13 M 1-methylimidazole, pH 8 (adjusted with HCl).
2. Positive-charged nylon membrane (Zeta-Probe blotting membrane, Bio-Rad).
3. 3 MM filter paper (Whatman, UK).
4. Semi-dry electrotransfer device (e.g., ATTO, Tokyo, Japan).

2.4. Detection of siRNAs by Northern Hybridization

1. Random primer DNA labeling kit (BcaBEST™ Labeling Kit, TaKaRa, Japan).
2. [α-^{32}P]dCTP (PerkinElmer).
3. MicroSpin™ G-25 column (GE Healthcare).

4. PerfectHyb^{TM} Plus (Sigma).
5. Heat-denatured salmon sperm DNA: Dissolve 100 mg of salmon sperm DNA in 10 mL of water (10 mg/mL). Autoclave the DNA solution and chill it immediately. Divide it into aliquots of about 1.0 mL in microtubes and store them at –20°C.
6. Hybridization oven (e.g., TAITEC, Tokyo, Japan) and bottle (35 mm in diameter × 150 mm in length).
8. X-ray film (Kodak BioMax MS-2).
9. Bio-imaging analyzer (BAS-1500, Fujifilm).

2.5. Isolation and Detection of Long dsRNAs by CF-11 Cellulose Chromatography and AGE

1. Liquid nitrogen.
2. 10X STE (saline–Tris–EDTA) buffer: 100 mM Tris–HCl, pH 8.0, 1.0 M NaCl, and 10 mM EDTA.
3. Extraction buffer for long dsRNA: 2X STE, 1% SDS, and 0.1% mercaptoethanol.
4. Phenol, chloroform, and isoamylalcohol (PCI) solution: water-saturated phenol with 0.1% 8-hydroxyquinoline: chloroform:isoamylalcohol = 25:24:1 (v/v/v).
5. 1X STE with 17.5% ethanol: 0.825 volume of 1X STE and 0.175 volume of absolute ethanol.
6. CF-11 cellulose powder (Whatman, UK).
7. Column for CF-11 cellulose chromatography, a glass column of about 1.5 cm diameter × 15 cm long or a disposable plastic column of about 1.0 cm diameter × 5 cm long.
8. Screw-capped tube.
9. Recombinant DNase I, RNase-free (TaKaRa, Japan).
10. S1 nuclease (TaKaRa, Japan).
11. 10X S1 nuclease buffer: 2.8 M NaCl, 300 mM sodium acetate, pH 4.6, and 10 mM $ZnSO_4$, which is attached with S1 nuclease.

2.6. Detection of Long dsRNAs by Native AGE and Northern Hybridization

1. Submerged and horizontal agarose gel electrophoresis chamber, gel size 120 mm (W) × 160 mm (L) (e.g., ATTO, Tokyo, Japan).
2. Denaturizing solution for long dsRNAs in agarose gel: 0.5 volume of formamide, 0.25 volume of formaldehyde solution (37%), and 0.25 volume of 1X TAE buffer.
3. 20X SSC: 3 M NaCl and 0.3 M sodium citrate.
4. Hybridization solution for long dsRNA or mRNA: 250 mM sodium phosphate buffer, pH 7.2, 1 mM EDTA, 7% SDS, and 1% BSA.

2.7. Detection of Long dsRNAs by Denaturing AGE and Northern Hybridization

1. 20X MOPS buffer.
2. Formaldehyde solution (37%, about 12 M).
3. Submerged and horizontal agarose gel electrophoresis chamber.

3. Methods

In 1999, Hamilton and Baulcombe first detected an siRNA from a small RNA fraction that was roughly isolated from total RNA (5). Here we report that a specific short dsRNA (siRNA) was directly detected from total RNA by northern hybridization. Applying the cross-linking method of small RNAs to a nylon membrane developed by Pall et al. increases the sensitivity of detection (9), thereby the amount of total RNA for hybridization can be reduced. A probe for detecting the specific siRNA can be made easily by a random primer DNA labeling kit. Therefore the method described here is simple and highly sensitive.

CF-11 cellulose column chromatography is one of the most useful procedures for isolating long dsRNAs from total nucleic acids, though it had been developed about 30 years ago (7, 8). Purified long dsRNAs can be detected by native AGE by staining with EtBr (10). By using either native or denaturing AGE, the dsRNA of interest can be detected specifically by northern hybridization.

3.1. Isolation of Total RNA from Plant Tissues

1. Isolate total RNA from plant tissues (e.g., about 50–100 mg of seedlings or leaves) by a Trizol reagent (Invitrogen) according to the manufacturer's protocol. Further purification to enrich small RNAs is not necessary for detecting the short dsRNA (siRNA) of interest by northern hybridization described here (*see* **Note 1**).
2. Suspend a pellet of total RNA with about 100 μL of DEPC-treated water (*see* **Note 2**).
3. Two microliters of total RNA solution is diluted to 50 times (100 μL) by water (*see* **Note 3**), and then measure its UV absorbance by a UV spectrophotometer. The peak of UV absorbance of RNA (nucleotide) solution is at around 260 nm.
4. Calculate the concentration of total RNA (*see* **Note 4**).
5. Add one-tenth volume of 3 M sodium acetate (pH 5.2) and 2.5 volume of 100% ethanol to the residual RNA solution (98 μL). This total RNA in 70% ethanol with 0.3 M

sodium acetate can be kept stably at –20°C for at least several months.

6. Quality of isolated total RNA, i.e., whether they are intact or fragmented, is examined by native mini-AGE (*see* **Note 5**). Centrifuge an aliquot of 70% ethanol containing an RNA precipitate (about 2 μg) to collect RNA, and then dry the precipitate (total RNA) in air.
7. Suspend total RNA with 16 μL of RNA sample buffer and 5 μL of water, incubate RNA samples at 65°C for 10 min, and immediately cool down on ice.
8. Set up a mini submerged AGE chamber and prepare a 1.2% mini agarose gel (about 50 mm × 60 mm) containing 1X TAE buffer and 200 ng/mL EtBr. Apply RNA samples on a mini gel and electrophorese them with 1X TAE buffer containing EtBr (200 ng/mL) at 100 V for about 30 min (or 50 V, 60 min).
9. Place the stained gel on a UV transilluminator and then photograph it. An example of the results is shown in **Fig. 9.1a**.
10. You can judge the quality (intact or fragmented) of isolated RNA from this photograph.
11. **Figure 9.1a** shows intact (lane 1) and degraded (lane 2) RNAs that correspond to distinct and smear rRNA bands,

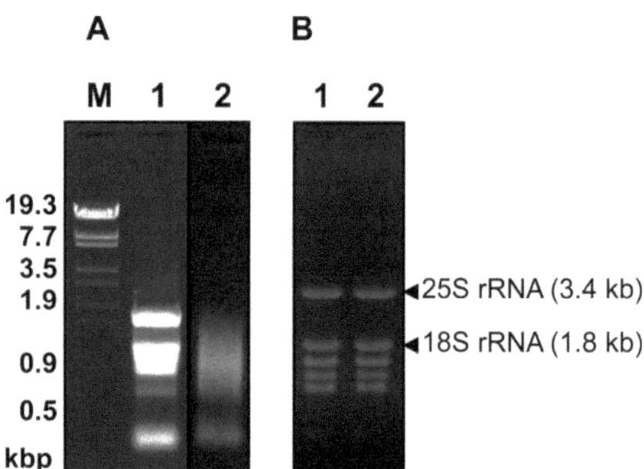

Fig. 9.1. Determination of quality of total RNA by AGE. (**a**) Total RNA was isolated from *Arabidopsis* leaves by Trizol reagent and electrophoresed by a native mini agarose gel of about 50 mm (W) × 60 mm (L). Intact (*lane 1*) and degraded (*lane 2*) total RNAs are shown. Ribosomal RNA (rRNA) bands of intact total RNA are distinct (*lane 1*), but those of degraded total RNA are smear (*lane 2*). *Lane M* is linear dsDNA markers. (**b**) An example of electrophoresis pattern of intact total RNAs by a denaturing agarose gel. Samples shown in *lanes 1* and *2* are intact total RNAs isolated from leaves of different *Arabidopsis* plants. The largest two bands indicated by *arrowheads* represent 25S rRNA and 18S rRNA of about 3.4 and 1.8 kb, respectively, which are encoded by a nuclear genome, and smaller bands represent chloroplast and mitochondrial rRNAs.

respectively. Furthermore, the precise concentration of intact total RNA can be estimated from band intensities of rRNAs stained by EtBr on the gel (*see* **Note 6**).

12. **Figure 9.1b** also shows an example of the electrophoresis pattern of high-quality (intact) total RNA isolated by this method, which was electrophoresed by denaturing agarose gel containing 1X MOPS buffer with 0.66 M formaldehyde and 200 ng/mL EtBr. Distinct bands of 25S and 18S rRNA (arrowheads in **Fig. 9.1b**) as well as chloroplast and mitochondrial rRNAs indicate that high-quality (intact) total RNA was obtained (*see* **Note 7**).

3.2. Denaturing PAGE for Short dsRNAs (siRNAs)

1. Set up a mini-slab gel electrophoresis chamber. The size of the gel is 90 mm (W) × 80 mm (L) × 1 mm (thick).
2. Prepare 10 mL of 17% denaturing polyacrylamide gel by mixing 4.2 g urea, 5.7 mL 30% acrylamide/bis solution, 1.0 mL 20X MOPS, 0.2 mL water, 80 μL 10% ammonium persulfate, and 10 μL TEMED. A gel should be polymerized within 30 min.
3. Prepare about 500 mL of a running buffer (2X MOPS) by diluting 20X MOPS with water (*see* **Note 8**).
4. Add the running buffer to the upper and lower chambers of the electrophoresis unit.
5. Centrifuge an aliquot of 70% ethanol containing RNA precipitates (about 2–5 μg of total RNA) to collect RNA and then dry precipitates (total RNA) in air. About 2–5 μg of total RNA is enough for detecting an siRNA derived from a hairpin-loop RNA, which is expressed from an artificially introduced knockdown (inverted repeat) construct with a strong promoter (e.g., 35S promoter) by this method.
6. Suspend total RNA with 15 μL of the formamide RNA solution, incubate RNA samples at 60°C for 10 min, and immediately cool down on ice.
7. Apply the RNA samples on the gel and electrophorese them at 10 mA (constant current) for about 1 h until the BPB dye reaches two-thirds of the gel.
8. After electrophoresis, stain the gel in 2X MOPS containing EtBr (200 ng/mL) for 5 min to detect rRNAs and tRNAs, which are used as loading controls. Destain the gel in 2X MOPS for 2 min.
9. Place the stained gel on a UV transilluminator and then photograph it. An example of the results is shown in **Fig. 9.2**.

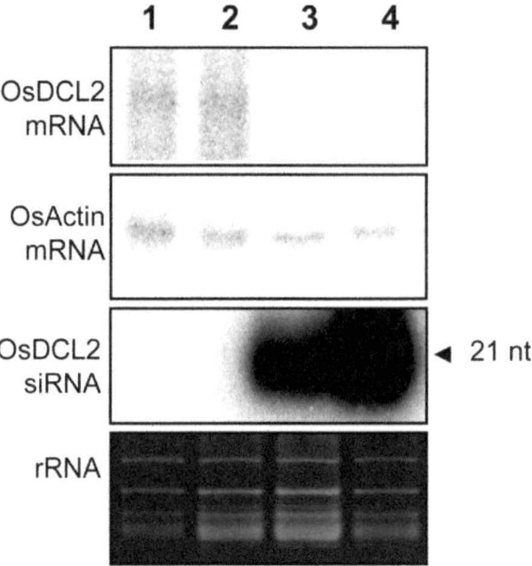

Fig. 9.2. Detection of short dsRNA (siRNA) derived from a hairpin-loop RNA by northern hybridization. Total RNA was isolated from wild-type rice plants (*lanes 1* and *2*) and *OsDCL2* (*Oryza sativa Dicer-like 2*)-knockdown (KD) rice plants (*lanes 3* and *4*). These two KD plants contain an inverted repeat construct for the *OsDCL2* gene. Total RNAs isolated from rice leaves were separated by denaturing AGE for mRNA or denaturing PAGE for short dsRNA (siRNA). The accumulation of *OsDCL2* mRNA and *OsDCL2* siRNAs was analyzed by northern hybridization with a ^{32}P-labeled DNA probe made by a random primer labeling kit. *OsActin1* (*Oryza sativa Actin 1*, hybridization) and rRNAs (EtBr staining) are loading controls for *OsDCL2* mRNA and siRNA, respectively. In KD plants (*lanes 3* and *4*) *OsDCL2* siRNA was detected but *OsDCL2* mRNA was not, indicating that RNAi was induced by a hairpin-loop RNA. Although only about 4 μg of total RNA was used in this experiment, siRNA was clearly detected in *lanes 3* and *4*.

3.3. Transfer and Cross-Linking of siRNAs to a Nylon Membrane

1. The essence of this cross-linking method was originally reported by Pall et al. (9), from which the following procedure was modified (11).
2. Set up a semi-dry electrotransfer device, which is also used to transfer proteins in western blotting.
3. Cut a positive-charged nylon membrane and six pieces of Whatman 3 MM paper to a slightly larger (or same) size as the gel.
4. Soak the membrane and 3 MM papers in distilled water for 5 min.
5. Assemble a sandwich for transfer as follows: from bottom (anode) to top (cathode); 3 MM paper (three pieces), membrane, gel, 3 MM paper (three pieces), cathode. Remove air bubbles by rolling the surface with a pipette.
6. Connect the transfer device to a power supply. Set a power supply 1.0 mA/1.0 cm^2 membrane (constant current) for 1 h at 4°C.
7. In each experiment, prepare a just sufficient volume of the cross-linking solution to saturate a single sheet of 3 MM filter paper that is slightly larger than the membrane.

8. After transfer, place the membrane immediately on the cross-linking solution-saturated 3 MM paper with the side onto which RNAs had been transferred facing up.
9. Wrap it in Saran wrap and incubate it at 60°C for about 1 h. Wash the membrane in distilled water to remove any residual cross-linking solution.
10. Air-dry the membrane on a new 3 MM paper.

3.4. Detection of siRNAs by Northern Hybridization

1. A hybridization oven and bottle are used for this method.
2. Synthesize the ^{32}P-labeled DNA probe for northern hybridization by a random primer DNA labeling kit with [α-^{32}P]dCTP according to the manufacturer's protocol (*see* **Note 9**).
3. Purify the ^{32}P-labeled DNA probe to remove free [α-^{32}P] dCTP by a MicroSpin™ G-25 column.
4. Prehybridization: Insert the membrane in a hybridization bottle, and add about 7 mL of a hybridization solution with heat-denatured salmon sperm DNA (0.1 mg/mL, final concentration) to the bottle. Prehybridization is carried out at 50°C for about 15–60 min.
5. After prehybridization, add the ^{32}P-labeled DNA probe to the hybridization solution, and incubate the membrane at 50°C for 16 h (overnight).
6. Wash the membrane four times in 2X SSC with 0.5% SDS at 50°C for 30 min.
7. Wrap the membrane in Saran wrap, and then analyze it by an imaging analyzer or by using an X-ray film (autoradiography). An example of the results is shown in **Fig. 9.2**.

3.5. Isolation and Detection of Long dsRNAs by CF-11 Cellulose Chromatography and AGE

1. Methods for isolating total nucleic acids from plant tissues and for purifying long dsRNAs from them by CF-11 cellulose chromatography (7, 8) are described in this section (*see* **Note 10**).
2. Suspend CF-11 cellulose powder in distilled water, wash it with distilled water several times to remove fine cellulose particles, wash it with 1X STE several times, and then finally equilibrate it with 1X STE with 17.5% ethanol. This cellulose slurry can be stored at 4°C for about 1 year.
3. Pulverize fresh or frozen tissue (e.g., leaf or seedling) in a mortar after freezing in liquid nitrogen, and then suspend it with the extraction buffer for long dsRNA. In the case of plant tissues, about 5–10 mL of the extraction buffer is usually used for 1 g of tissues.
4. Add an equal volume of the PCI solution to the homogenate. Mix vigorously by a vortex mixer, and

centrifuge it at about 10,000×*g* (10,000 rpm) for 20 min at room temperature.

5. After centrifugation, place the aqueous phase into a new tube. Do not take white precipitates at the boundary between aqueous and chloroform phases, because they contain denatured proteins. Add an equal volume of the PCI solution to the aqueous phase, mix vigorously, and centrifuge it again. Place the aqueous phase into a new tube again. Two extractions with PCI solution are usually enough for isolating long dsRNAs from plant tissues, such as leaves or seedlings (*see* **Note 11**).
6. Add 2 volumes of 100% ethanol to the aqueous phase and store it at –20°C for more than 1 h. Centrifuge it at about 10,000×*g* (10,000 rpm) for 10 min and discard the supernatant. The obtained pellet contains total nucleic acids.
7. Suspend the pellet completely with 1X STE containing 17.5% ethanol in a screw-capped tube. Incubate it at 60°C for 10 min, and transfer it on ice immediately to denature the secondary structures of ssRNAs, especially rRNAs (*see* **Note 12**). If precipitates appear, they should be removed by centrifugation at about 10,000×*g* (10,000 rpm) for 20 min at 4°C.
8. Add about 0.5 volume of CF-11 cellulose slurry to the heat-denatured nucleic acid solution and mix gently at 4°C for 30 min. In this step cellulose adsorbs dsRNAs.
9. Pack the mixture of cellulose with nucleic acids into a column, whose size depends on the volume of cellulose.
10. Wash the cellulose thoroughly with 1X STE with 17.5% ethanol until most DNAs and ssRNAs are eluted (*see* **Note 13**).
11. Elute dsRNAs with 1X STE and collect them.
12. Add 1/10 volume of 3 M sodium acetate (pH 5.2) and 3 volumes of 100% ethanol to the eluate containing dsRNAs and store it at –20°C for more than 2 h (or overnight) (*see* **Note 14**). Centrifuge it at about 10,000×*g* (10,000 rpm) for 10 min and discard the supernatant. The pellet contains dsRNAs.
13. Suspend the pellet with 2X STE (*see* **Note 15**).
14. Analyze isolated dsRNAs by electrophoresis with a 1.0% agarose gel containing 1X TAE buffer and 200 ng/mL EtBr (native AGE). Linear dsDNA markers can be used as a size marker for long dsRNAs, because linear dsRNAs move like linear dsDNAs in native AGE. A long dsRNA looks like a linear dsDNA by staining with EtBr. An example of the results is shown in **Fig. 9.3**.

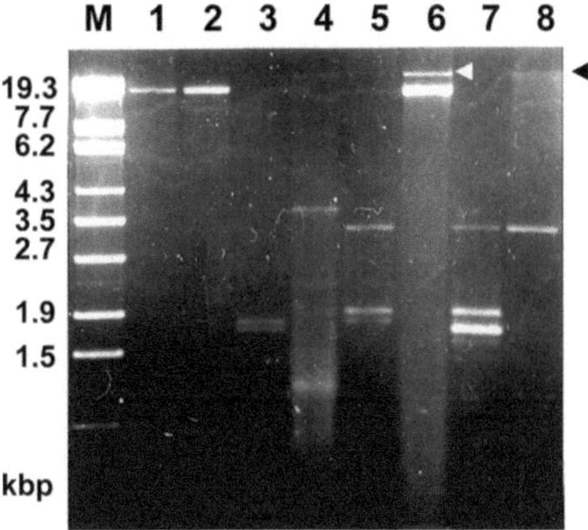

Fig. 9.3. Isolation and detection of long dsRNAs from various symptomless plants by CF-11 cellulose chromatography and native AGE. By using CF-11 cellulose chromatography long dsRNAs were isolated from rice (*Oryza sativa, lane 1*), seagrass (*Zostera marina, lane 2*), radish (*Raphanus sativus, lane 3*), watercress (*Rorip nasturtiumaquaticum, lane 4*), spinach (*Spinacia oleracea, lane 5*), Malabar spinach (*Basella alba, lane 6*), crown daisy (*Chrysanthemum coronarium, lane 7*), and dandelion (*Taraxacum platycarpum, lane 8*). They were separated by native AGE at 20 V for about 18 h and stained by EtBr. *Lane M* is linear dsDNA markers. Lengths of dsRNAs are comparable with linear dsDNA markers and are from about 1.2 to 15 kbp. *Arrowheads* in *lanes 6* and *8* may be DNA that was not removed by CF-11 cellulose chromatography. (Reproduced from ref. 10 with permission from Springer.)

15. If the dsRNA fraction contains a small amount of ssRNAs with secondary structure (rRNAs) and DNAs (*see* **Note 16**), further purification by treatments with RNase-free DNase I and S1 nuclease is effective to yield purer dsRNAs (*see* **Note 17**).

3.6. Detection of Long dsRNAs by Native AGE and Northern Hybridization

1. A cDNA clone is necessary for detecting a long dsRNA by northern hybridization. Two electrophoresis methods, denaturing AGE and native AGE, are applicable for detecting a long dsRNA by northern hybridization.
2. If you have purified dsRNAs, a combination of native AGE and northern hybridization may be better than combining denaturing AGE and northern hybridization, because you can detect long dsRNAs on native AGE (**Figs. 9.3** and **9.4a**) but not on denaturing AGE by staining with EtBr.
3. Set up a submerged and horizontal AGE chamber and prepare a 1.0% agarose gel (about 120 mm (W) × 160 mm (L)) with 1X TAE buffer containing EtBr (200 ng/mL). Electrophorese dsRNAs with 1X TAE buffer containing EtBr (200 ng/mL) at 50 V (constant voltage) for about 6 h (or at 20 V for about 18 h) (*see* **Note 18**).
4. Place the gel on a UV transilluminator, and photograph it to measure the mobility of dsRNAs in comparison with linear

dsDNA markers as described in **Section 3.5**, Step 14. An example is shown in **Figs. 9.3** and **9.4a**.

5. Transfer the gel to a plastic container and incubate it in a denaturizing solution for 50 min at 60°C (12, 13) (*see* **Note 19**).
6. Incubate the gel in 20X SSC for 15 min at room temperature.
7. Transfer the denatured dsRNAs to a nylon membrane by a capillary transfer method with 20X SSC.
8. The standard northern hybridization method for detecting mRNA is applicable for detecting the denatured dsRNAs on a nylon membrane.
9. Synthesize a probe for hybridization from a cDNA clone by a random primer DNA labeling kit as described in **Section 3.4**, Step 2.
10. Prehybridization and hybridization are carried out according to the manufacturer's protocol of a nylon membrane. About 5–10 mL of the hybridization solution with heat-denatured salmon sperm DNA (final concentration,

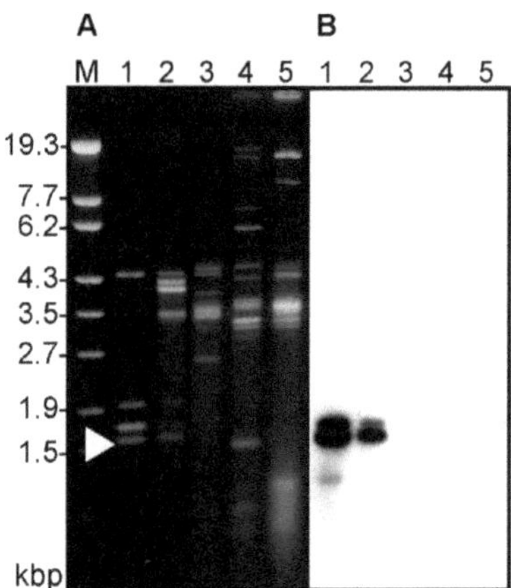

Fig. 9.4. Determination of a specific dsRNA segment by northern hybridization with native AGE. Relationship of the smallest dsRNA segment from green alga *Bryopsis cinicola* (*lane 1*, *white arrowhead*) with dsRNAs from a variety of green algae was analyzed by native AGE and northern hybridization. (**a**) Various dsRNAs were purified from a variety of green algae; *B. cinicola* (*lane 1*), *B. maxima* (*lane 2*), *Codium cylindricum* (*lane 3*), *C. latum* (*lane 4*), and *C. pugniformis* (*lane 5*). They were separated by native AGE and stained by EtBr. *Lane M* is linear dsDNA markers. (**b**) Northern hybridization of algal dsRNAs with a cDNA probe derived from the smallest dsRNA segment from *B. cinicola* (lane 1, *white arrowhead*). The cDNA probe hybridized with the second smallest dsRNA segment from *B. cinicola* (*lane 1*) and the two smallest dsRNA segments from *B. maxima* (*lane 2*) as well as the smallest dsRNA segment from *B. cinicola* (*lane 1*), indicating that these four dsRNA segments from two closely related green algae are similar to each other. (Reproduced from ref. 13 with permission from Springer.)

0.1 mg/mL) is used for prehybridization, which is carried out at 65°C for about 15–60 min.

11. After prehybridization, add the heat-denatured ^{32}P-labeled DNA probe to the hybridization solution and incubate the membrane at 65°C for 16 h (overnight).
12. Wash the membrane twice with 40 mM sodium phosphate buffer (pH 7.2) and 1 mM EDTA containing 5% SDS at 65°C for 30–60 min and twice with the same buffer containing 1% SDS at 65°C for 30–60 min.
13. Analyze it by an imaging analyzer or by using an X-ray film. An example of the results is shown in **Fig. 9.4b**. You can determine which dsRNA segment the cDNA clone originated from.

3.7. Detection of Long dsRNAs by Denaturing AGE and Northern Hybridization

1. Northern hybridization by using denaturing AGE for detecting a long dsRNA is the same as the standard northern hybridization method for mRNA detection.
2. Either total RNA or purified dsRNA is applicable. A cDNA clone derived from the dsRNA of interest is necessary.
3. Suspend total RNA or purified dsRNA with the RNA sample buffer, incubate RNA samples at 65°C for 10 min, and immediately cool down on ice.
4. Set up a submerged and horizontal AGE chamber, and prepare a 1.2% agarose gel (about 120 mm (W) × 160 mm (L)) with 1X MOPS buffer, 0.66 M formaldehyde, and 200 ng/mL EtBr. Electrophorese dsRNAs with 1X MOPS buffer containing EtBr (200 ng/mL) at 50 V for about 5 h.
5. Transfer denatured dsRNAs to a nylon membrane by the capillary transfer method with 20X SSC.
6. The standard northern hybridization method for detecting mRNA is applicable for detecting denatured dsRNA on the nylon membrane. Hybridization is carried out as described in **Section 3.6**, Steps 9–13.

4. Notes

1. Some small RNA isolation kits can be commercially available for short dsRNAs (siRNAs) now, such as *mir*Vana^{TM} miRNA isolation kit (Ambion).
2. When 50–100 mg of plant tissue is used for isolating total RNA, about 100 μL of DEPC-treated water is used for dissolving the total RNA.

3. The volume of RNA solution depends on the cuvette of the UV spectrophotometer.
4. Concentration of RNA is the value of absorbance at 260 nm × 45 μg/mL.
5. This step is important, because the quality of isolated RNA is a critical determinant for the success of downstream experiments. It is well known that RNA is a fragile molecule, and it is easily fragmented, enzymatically or chemically.
6. Concentration of RNA can be calculated from the UV absorbance of RNA solution, but this RNA solution usually contains not only intact RNAs but also fragmented (degraded) RNAs and oligonucleotides. Therefore, more precise concentration of intact RNAs should be judged from the band intensities of rRNAs on AGE (*see* **Fig. 9.1**).
7. These total RNAs separated by denaturing AGE can be transferred to a nylon membrane by the capillary transfer method, and then the membrane can be used to detect a specific mRNA by northern hybridization.
8. Not Tris–borate buffer but 1X MOPS buffer was used as a running buffer in an original paper by Pall et al. (9). However, when 1X MOPS was used, the electrophoresis pattern of total RNA was distorted. We recommend 2X MOPS as a running buffer.
9. This probe is also usable for detecting mRNA by northern hybridization.
10. Expensive equipment for CF-11 cellulose column chromatography is not necessary as described in **Section 2.5**.
11. If denatured proteins at the boundary are very visible, re-extraction with PCI is recommended.
12. rRNAs are the most abundant RNA species (more than 80%) among total RNAs, and they form secondary structures containing many stem-loop regions, i.e., dsRNA regions. Therefore, it is difficult to remove rRNAs completely from the dsRNA fraction.
13. You can determine by monitoring the UV absorbance at 260 nm of eluate whether most DNAs and ssRNAs are eluted.
14. Because the concentration of dsRNA may be low, long incubation at –20°C for precipitating dsRNA is recommended for high recovery of dsRNA. Otherwise you can use glycogen (Roche Diagnostics) as a carrier for the precipitation of nucleic acids by ethanol.
15. dsRNAs are relatively stable in high-salt buffer, such as 2X STE (0.2 M NaCl), but are unstable in low-salt

buffer, because in low-salt buffer dsRNAs partially expose the single-stranded region, which is accessible by ssRNA-specific ribonucleases, such as RNase A. ssRNA-specific ribonucleases are ubiquitous and major in every cell, but dsRNA-specific ribonucleases are minor, indeed only RNase III family members are known as dsRNA-specific ribonucleases so far (14). Consequently dsRNAs are more stable than ssRNAs.

16. The highest molecular weight bands (arrowheads) in lanes 6 and 8 of **Fig. 9.3** must be DNA that was not removed by CF-11 cellulose chromatography.
17. DNase I and S1 nuclease can be used simultaneously in 1X S1 nuclease buffer.
18. Better resolution of nucleic acids in AGE is obtained from the conditions with lower voltage and longer running time (e.g., 20 V for about 18 h) than those with higher voltage and shorter running time (e.g., 100 V for about 3 h).
19. Vapor from a denaturizing solution is toxic, because it contains formaldehyde. A plastic container should be sealed and the incubation should be carried out in a draft chamber.

Acknowledgments

The authors would like to thank Dr. Hiroaki Kodama, Chiba University, for his advice. This research was supported in part by Grant-in-Aids for Scientific Research to T.F. and H.M. from the Ministry of Education, Science, Sports, Culture and Technology of Japan.

References

1. Fire, A., Xu, S., Montgomery, M. K., Kostas, S. A., Driver, S. E., and Mello, C. C. (1998) Potent and specific genetic interference by double-stranded RNA in *Caenorhabditis elegans. Nature* **391**, 806–811.
2. Napoli, C., Lemieux, C., and Jorgensen, R. (1990) Introduction of a chimeric chalcone synthase gene into petunia results in reversible co-suppression of homologous genes in trans. *Plant Cell* **2**, 279–289.
3. van der Krol, A. R., Mur, L. A., Beld, M., Mol, J. N., and Stuitje, A. R. (1990) Flavonoid genes in petunia: addition of a limited number of gene copies may lead to a suppression of gene expression. *Plant Cell* **2**, 291–299.
4. Ding, S. W., and Voinnet, O. (2007) Antiviral immunity directed by small RNAs. *Cell* **130**, 413–426.
5. Hamilton, A. J. and Baulcombe, D. C. (1999) A species of small antisense RNA in posttranscriptional gene silencing in plants. *Science* **286**, 950–952.
6. Brown, G. G. and Finnegan, P. M. (1989) RNA plasmids. *Int. Rev. Cytol.* **117**, 1–56.
7. Morris, T. J. and Dodds, J. A. (1979) Isolation and analysis of double-stranded RNA from virus-infected plant and fungal tissue. *Phytopathology* **69**, 854–858.

8. Schuster, A. M. and Sisco, P. H. (1986) Isolation and characterization of single-stranded and double-stranded RNAs in mitochondria. *Methods Enzymol.* **118**, 497–507.
9. Pall, G. S., Codony-Servat, C., Byrne, J., Ritchie, L., and Hamilton, A. (2007) Carbodiimide-mediated cross-linking of RNA to nylon membranes improves the detection of siRNA, miRNA and piRNA by northern blot. *Nucleic Acids Res.* **35**, e60.
10. Fukuhara, T., Koga, R., Aoki, N., Yuki, C., Yamamoto, N., Oyama, N. et al. (2006) The wide distribution of endornaviruses, large double-stranded RNA replicons with plasmid-like properties. *Arch. Virol.* **151**, 995–1002.
11. Urayama, S., Moriyama, H., Aoki, N., Nakazawa, Y., Okada, R., Kiyota, E. et al. (2010) Knockdown of *OsDCL2* in rice negatively affects maintenance of the endogenous dsRNA virus, *Oryza sativa endornavirus. Plant Cell Physiol.* **51**, 58–67.
12. Zabalgogeazcoa, I. A., Cox-Fostre, D. C., and Gildow, F. E. (1993) Pedigree analysis of the transmission of a double-stranded RNA in barley cultivars. *Plant Sci.* **91**, 45–53.
13. Koga, R., Horiuchi, H., and Fukuhara, T. (2003) Double-stranded RNA replicons associated with chloroplasts of a green alga, *Bryopsis cinicola. Plant Mol. Biol.* **51**, 991–999.
14. Nicholson, A. W. (2003) The ribonuclease III family: forms and functions in RNA maturation, decay, and gene silencing. In *RNAi: A Guide to Gene Silencing* (ed. Hannon, G. J.), Cold Spring Harbor Press, Cold Spring Harbor, NY, pp. 149–174.

Chapter 10

Quantitative Stem-Loop RT-PCR for Detection of MicroRNAs

Erika Varkonyi-Gasic and Roger P. Hellens

Abstract

Plant microRNAs (miRNAs) are a class of endogenous small RNAs that are essential for plant development and survival. They arise from larger precursor RNAs with a characteristic hairpin structure and regulate gene activity by targeting mRNA transcripts for cleavage or translational repression. Efficient and reliable detection and quantification of miRNA expression has become an essential step in understanding their specific roles. The expression levels of miRNAs can vary dramatically between samples and they often escape detection by conventional technologies such as cloning, northern hybridization and microarray analysis. The stem-loop RT-PCR method described here is designed to detect and quantify mature miRNAs in a fast, specific, accurate and reliable manner. First, a miRNA-specific stem-loop RT primer is hybridized to the miRNA and then reverse transcribed. Next, the RT product is amplified and monitored in real time using a miRNA-specific forward primer and the universal reverse primer. This method enables miRNA expression profiling from as little as 10 pg of total RNA and is suitable for high-throughput miRNA expression analysis.

Key words: miRNA, qPCR, stem-loop RT, SYBR Green, UPL.

1. Introduction

Small RNAs are 19–25 nucleotide long non-coding RNA molecules that include short interfering RNAs (siRNAs), mediators of post-transcriptional and transcriptional gene silencing (1), and microRNAs (miRNAs), essential for processes ranging from developmental patterning to stress responses (2–5). Both classes of small RNAs are derived from longer precursor RNA molecules. Long double-stranded RNA precursors give rise to siRNAs, and single-stranded RNA precursors with a characteristic hairpin secondary structure give rise to miRNAs (reviewed in ref. 2). Both classes of small RNAs repress gene

H. Kodama, A. Komamine (eds.), *RNAi and Plant Gene Function Analysis*, Methods in Molecular Biology 744, DOI 10.1007/978-1-61779-123-9_10, © Springer Science+Business Media, LLC 2011

activity by targeting perfect complementary sequences (siRNAs) or near-perfect complementary sequences (miRNA) in target mRNAs to guide cleavage and translational repression (6–10) or on DNA to guide chromatin modification (11). The majority of plant miRNA targets are transcription factors essential for development (12, 13) and genes involved in stress responses (14, 15). Expression of a miRNA may result in a complete removal of the corresponding target mRNA thus enabling cell-fate changes (13, 16, 17) or in a reduction in the level of the target mRNA and co-expression of both the miRNA and the target mRNA in the same tissue (18, 19). In addition, some miRNAs are transported in the phloem and may work in a non-cell-autonomous manner (20, 21), in contrast to miRNAs with demonstrated cell-autonomous expression and effects (22, 23).

The important role that miRNAs play in regulation of gene expression and the complexity of their mode of action demonstrate that reliable detection and quantification of miRNA expression is essential for understanding of the miRNA-mediated gene regulation. Conventional technologies such as cloning, northern hybridization and microarray analysis are widely used but are time-consuming, require relatively large amounts of RNA and are not sensitive enough to detect less-abundant miRNAs. Furthermore, small RNA fraction in plants is highly complex, which further affects detection methods such as cloning and microarray hybridization (12, 24). In contrast, reverse transcription polymerase chain reaction (RT-PCR) detection methods are very sensitive, specific, fast and require small amounts of RNA for analysis; however, the standard RT-PCR is not amenable for detection of RNA with an average length of 22 nt. Different methods were developed to design assays for detection of miRNAs, including RNA tailing to extend the template length and various primer designs to incorporate additional nucleotides during the reverse transcription step. Assays based on linear primer design often lacked specificity, and additional improvements were required, e.g. incorporation of locked nucleic acid (LNA®) nucleosides used in some commercially available miRNA detection kits (Exiqon A/S, Vedbaek, Denmark).

Stem-loop reverse transcription primers were shown to provide better specificity and sensitivity than linear primers (25), and a pulsed reverse transcription (RT) reaction, using short cycles of incubation at gradually increasing annealing temperatures to ensure correct pairing, further increased the sensitivity of miRNA detection (26). These features were utilized in a two-step miRNA detection method (**Fig. 10.1**). First, the stem-loop RT primer is hybridized to a specific miRNA molecule and then reverse transcribed in a pulsed RT reaction. Next, the RT product is amplified using a miRNA-specific forward primer and the universal reverse primer. The product can be visualized by gel electrophoresis upon

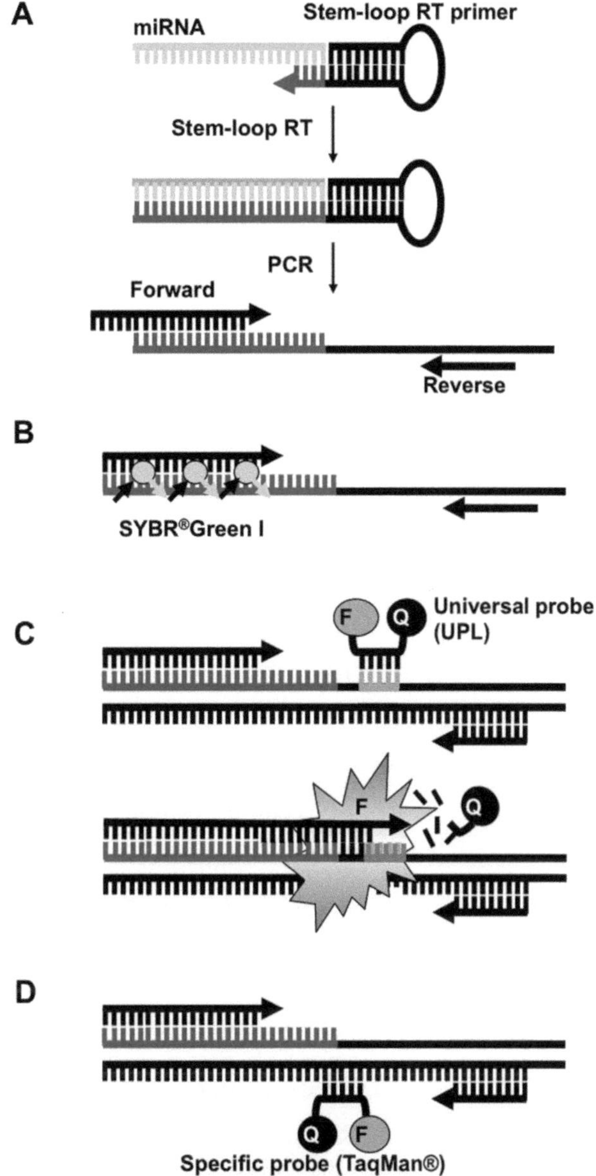

Fig. 10.1. Schematic showing the stem-loop RT-PCR for detection of microRNAs. (**a**) A stem-loop RT primer binds to the 3′ portion of the miRNA, initiating reverse transcription. The RT product is amplified using a miRNA-specific forward primer and the universal reverse primer. (**b**) The quantification is achieved through SYBR Green I incorporation during amplification. (**c**) The quantification is achieved by the fluorescence generated upon cleavage of the UPL probe, a hydrolysis probe of eight nucleotides including one locked nucleic acid (LNA) that increases binding specificity. This probe is designed to hybridize to a region within the amplicon and is dual labelled with a reporter dye and a quenching dye. The quenching dye is suppressing the fluorescence of the reporter dye while in close proximity. Once the probe is degraded by the exonuclease activity of Taq polymerase, the fluorescence of the reporter increases at a rate proportional to the amplification level. (**d**) A miRNA-specific TaqMan probe can be used in the same manner as the universal UPL probe.

a set number of PCR cycles or monitored in real time using quantitative PCR (qPCR). Quantification of miRNA expression can be performed using the SYBR® Green detection technology or a hydrolysis probe technology for increased specificity, such as the Universal ProbeLibrary (UPL; Roche Diagnostics, Mannheim, Germany) or TaqMan® (Applied Biosystems, Foster City, CA). In addition to expression analysis of endogenous miRNAs, this method is amenable for detection and quantification of other small RNAs, including artificial miRNAs and synthetic siRNAs.

2. Materials

2.1. Plant Material and RNA

1. Plant tissue collected into liquid nitrogen and handled according to standard practices to prevent degradation of RNA.
2. Total RNA extracted using standard methods and handled according to standard laboratory practices to avoid RNase contamination. Avoid RNA purification using RNA-binding glass fibre filters (*see* **Note 1**). Prior to reverse transcription, RNA should be quantified and evaluated (*see* **Note 2**).

2.2. Stem-Loop Pulsed RT

1. Stem-loop RT primers. Prepare 100 μM stocks for long-term storage and 1 μM dilutions for immediate use.
2. 10 mM dNTP mix. Prepare by mixing dATP, dCTP, dGTP and dCTP stock solutions, aliquot out and store at –20°C.
3. Reverse transcriptase, e.g. SuperScript® III RT, 200 units/μL that is supplied with the First-Strand buffer for cDNA synthesis and 0.1 M DTT (Invitrogen).
4. RNase inhibitor, e.g. RNaseOUT™, 40 U/μL (Invitrogen).
5. Nuclease-free water.

2.3. miRNA SYBR Green qPCR Assay

1. LightCycler FastStart SYBR Green I master mix (Roche Diagnostics), prepared according to the manufacturer's instructions.
2. Universal reverse primer. Prepare 100 μM stock for long-term storage and 10 μM dilution for immediate use.
3. Forward miRNA-specific primer. Prepare 100 μM stock for long-term storage and 10 μM dilution for immediate use.
4. 10 mM dNTP mix as above.
5. Nuclease-free water.

2.4. miRNA Hydrolysis Probe qPCR Assay

1. LightCycler TaqMan master mix (Roche Diagnostics) prepared according to the manufacturer's instructions.
2. UPL probe #21 prepared as 10 μM stock (Roche Diagnostics) (*see* **Note 3**).
3. Universal reverse oligo. Prepare 100 μM stock for long-term storage and 10 μM dilution for immediate use.
4. Forward miRNA-specific oligonucleotide. Prepare 100 μM stock for long-term storage and 10 μM dilution for immediate use.
5. 10 mM dNTP mix as above.
6. Nuclease-free water.

2.5. Equipment

1. Standard laboratory equipment for isolation of RNA (fume hood, centrifuge, tubes, pipettes and tips).
2. Spectrophotometer for quantification of RNA, e.g. NanoDrop™ ND-1000 Spectrophotometer (NanoDrop Technologies, Wilmington, DE) (*see* **Note 2**).
3. Standard gel electrophoresis equipment (casting trays, gel tanks, power supply, UV transilluminator).
4. Thermal cycler for pulsed reverse transcription, e.g. Mastercycler® (Eppendorf, Hamburg, Germany).
5. A real-time thermal cycler, e.g. LightCycler® range (Roche Diagnostics).

3. Methods

3.1. Primer Design

The primers are designed according to Chen et al. (25) with some modifications (27) (**Fig. 10.2**). The stem-loop RT primers have a universal backbone and a specific extension. The universal backbone sequence is designed to form a stem-loop structure because of the complementarity between the nucleotides in the 5′- and 3′-end. It includes the reverse complement of the UPL probe #21 and the universal reverse primer site in the loop region. The specificity of a stem-loop RT primer to an individual miRNA is conferred by a six-nucleotide extension at the 3′-end. This extension is a reverse complement of the last six nucleotides at the 3′-end of the miRNA.

Forward primers are specific to the miRNA sequence but exclude the last six nucleotides at the 3′-end of the miRNA. A 5′ extension of five to seven nucleotides is added to each forward primer to increase the length and the melting temperature; these sequences were chosen randomly and are relatively GC rich, bringing the GC content of the forward primer to 50–60%.

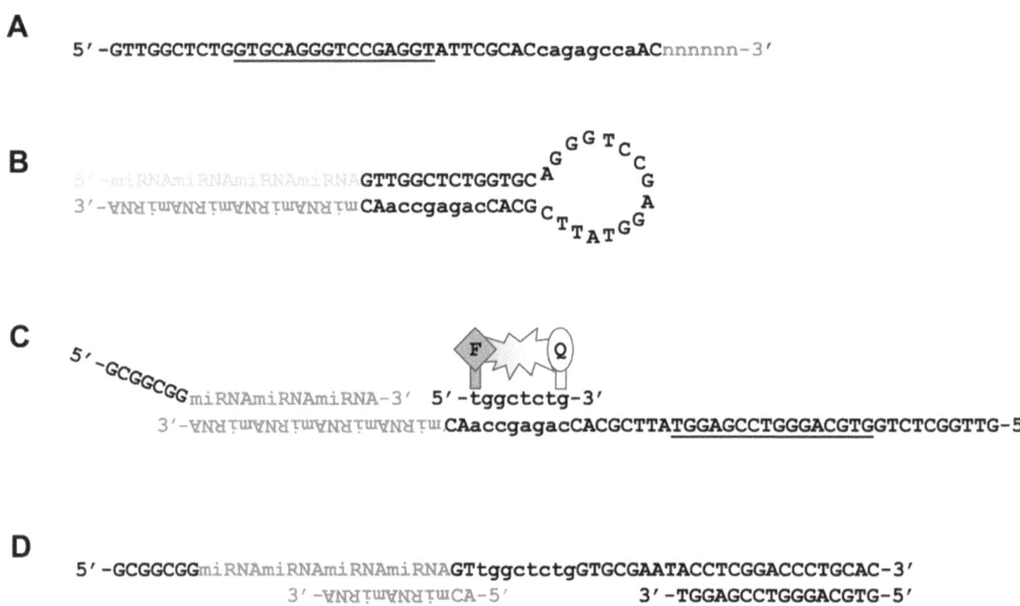

Fig. 10.2. Schematic showing the primer design for stem-loop RT-PCR for detection of microRNAs. (**a**) The stem-loop RT primers have a universal backbone and a miRNA-specific extension. The specificity of a stem-loop RT primer to an individual miRNA is conferred by a six-nucleotide extension at the 3′-end (nnnnnn) which is a reverse complement of the last six nucleotides at the 3′-end of the miRNA. The backbone includes the reverse complement of the UPL probe #21 (in *lower case*) and the universal reverse primer site (*underlined*). (**b**) The backbone sequence can form a stem-loop structure because of the complementarity between the nucleotides in the 5′- and 3′-end. Reverse transcription is initiated upon annealing to the six nucleotides at the 3′-end of the miRNA. (**c**) Forward primers are specific to the miRNA sequence but exclude the six nucleotides at the 3′-end of the miRNA. A 5′ extension of five to seven nucleotides is added to each forward primer to increase the length and the melting temperature. The 5′ extension sequences were chosen randomly and are relatively GC rich, bringing the GC content of the forward primer to 50–60%. The UPL probe #21 sequence can hybridize to the DNA but is removed by the exonuclease activity of Taq polymerase, resulting in the detectable fluorescence of the reporter dye. (**d**) A miRNA-specific TaqMan probe can be designed to distinguish between highly homologous targets.

3.2. Stem-Loop Pulsed RT

The most reproducible results are obtained with 1–10 ng of total RNA per reaction, but moderately and highly abundant miRNAs can be detected from as little as 10 pg of total RNA. The protocol is designed to quantify expression of a specific miRNA in a large number of biological samples or expression of a large number of miRNAs in one sample. If testing expression of one miRNA in many RNA samples, prepare a ‘no RNA’ master mix; if testing expression of many different miRNAs in one RNA sample, prepare a ‘no RT primer’ master mix. If handling large number of samples keep the reactions on ice and work in the cold room.

1. Denature the appropriate 1 μM stem-loop RT primer by heating to 65°C for 5 min.
2. Incubate on ice for 2 min.

3. Centrifuge briefly to bring solution to the bottom of the tube. Incubate on ice.
4. Prepare the 'no RNA' or 'no RT' master mix by scaling the volumes for a single RT reaction to the desired number of RT reactions. At least three replicates per RT reaction are recommended. It is a good practice to prepare additional 10% of master mix to cover pipetting errors. A single 'no RNA' reaction mix is prepared by adding the following components to a nuclease-free microcentrifuge tube:

   0.5 μL 10 mM dNTP mix,

   11.15 μL nuclease-free water,

   1 μL of denatured stem-loop RT primer,

   4 μL 5X First-Strand buffer,

   2 μL 0.1 M DTT,

   0.1 μL RNaseOUT (40 U/μL),

   0.25 μL SuperScript III RT (200 U/μL).
5. A single 'no RT' reaction mix is prepared as above by adding 1 μL of appropriate RNA template (*see* **Note 4**) instead of 1 μL denatured stem-loop RT primer.
6. Mix gently and centrifuge to bring solution to the bottom of the tube.
7. Assemble the RT reaction by aliquoting 19 μL of the 'no RNA' master mix and adding 1 μL RNA template (*see* **Note 4**) or 19 μL of the 'no RT' master mix and adding 1 μL of appropriate denatured stem-loop RT primer.
8. Mix gently and centrifuge to bring solution to the bottom of the tube.
9. Prepare the 'minus RT' controls by omitting reverse transcriptase from the reactions and the 'no template' controls by adding nuclease-free water in place of RNA.
10. Load thermal cycler and incubate for 30 min at 16°C, followed by pulsed RT of 60 cycles at 30°C for 30 s, 42°C for 30 s and 50°C for 1 s.
11. Incubate at 85°C for 5 min to inactivate the reverse transcriptase.

3.3. miRNA SYBR Green qPCR Assay

SYBR Green I assay provides good specificity if the number of PCR cycles is limited to 35 to minimize non-specific amplification. This number of cycles is sufficient for quantification of moderately abundant miRNAs.

1. Prepare 5X LightCycler FastStart SYBR Green I master mix according to the manufacturer's instructions.
2. Prepare a PCR master mix by scaling the volumes listed below to the desired number of amplification reactions.

Prepare additional 10% of master mix to cover pipetting errors. For a single reaction, add the following components to a nuclease-free microcentrifuge tube:

12 μL nuclease-free water,

4 μL SYBR Green I master mix,

1 μL 10 μM forward (miRNA-specific) primer,

1 μL 10 μM reverse (universal) primer.

3. Mix gently and centrifuge to bring solution to the bottom of the tube.
4. Store in cooling block or on ice.
5. Pipette 18 μL master mix into the multi-well plate or capillary.
6. Add 2 μL RT product.
7. Seal the multi-well plate or capillary and place into the LightCycler instrument.
8. Incubate samples at 95°C for 5 min, followed by 35–40 cycles of 95°C for 5 s and 60°C for 10 s.
9. For melting curve analysis, denature samples at 95°C, then cool to 65°C at 20°C/s. Collect fluorescence signals at 530 nm wavelength continuously from 65 to 95°C at 0.2°C/s.

3.4. miRNA Hydrolysis Probe qPCR Assay

For miRNA sequences that are expressed at low levels or when a particular set of primers produces background amplification, a hydrolysis probe assay provides higher specificity. The protocol for UPL probe assay is presented here that utilizes a single, universal hydrolysis probe, UPL probe #21 (Roche Diagnostics) to distinguish between specific amplicons and primer dimers. MicroRNA assay kits for detection of individual miRNAs using individual miRNA-specific TaqMan probes are available commercially (Applied Biosystems) (*see* **Note 3**).

1. Prepare 5X LightCycler TaqMan master mix (Roche Diagnostics) according to the manufacturer's instructions.
2. Prepare a PCR master mix by scaling the volumes listed below to the desired number of amplification reactions. Prepare additional 10% of master mix to cover pipetting errors. For a single reaction, add the following components to a nuclease-free microcentrifuge tube:

11.8 μL nuclease-free water,

4 μL TaqMan master mix,

1 μL 10 μM forward (miRNA-specific) primer,

1 μL 10 μM reverse (universal) primer,

0.2 μL 10 μM UPL probe #21.

3. Mix gently and centrifuge to bring solution to the bottom of the tube.
4. Store in cooling block or on ice.
5. Pipette 18 μL master mix into the multi-well plate or capillary.
6. Add 2 μL RT product.
7. Seal the multi-well plate or capillary and place into the LightCycler instrument.
8. Incubate samples at 95°C for 5 min, followed by 35–45 cycles of 95°C for 5 s and 60°C for 10 s.

3.5. Data Analysis

General instructions for data analysis are provided. For detailed instructions, refer to the appropriate instrument and software user manual. The quantitative PCR data can be analysed and presented as absolute or relative values. Relative quantification is the preferred method because it takes into account the potential errors due to variation in RNA input and RT efficiency (*see* **Note 5**). The most accurate method to correct these potential errors is normalization to endogenous control genes (*see* **Note 6**).

1. Perform melting curve analysis. This analysis is done after the SYBR Green I assay to determine that each of the primer pairs amplified a single predominant product with a distinct melting temperature (T_m). Follow the instrument user manual for instructions for melting curve analysis and T_m calling. If a single melting peak is observed for a particular primer pair, it is likely that a single product with a distinct T_m was amplified. Evaluate by gel electrophoresis (*see* **Note 7**).
2. Perform relative quantification. Relative quantification analysis compares two ratios: the ratio of the target gene to a reference gene sequence in an unknown sample is compared with the ratio of the same two sequences in a standard sample called a calibrator.
3. To perform relative quantification with an external standard, prepare standard curves for the target and reference genes by serial dilutions of external standards with a known copy number (*see* **Note 8**). Use at least three points or one point per log of concentration, whichever is greater. Always use a 'no template' control.
4. Prepare master mix and perform PCR as described earlier. Use at least three replicates per standard dilution and 'no template' control.
5. Follow the instrument user manual for instructions for the standard method that will automatically calculate and display the amplification curves and the standard curve, crossing points, calculated concentrations and statistics for replicates.
6. Save as standard curve object.

7. Perform relative quantification, calibrator normalized, without efficiency correction: select relative quantification – monocolor analysis, assign a 'target calibrator' and a 'reference calibrator' sample, assign appropriate pairs of target and reference samples and perform analysis following the instrument user manual (*see* **Note 9**).
8. Perform efficiency correction by applying standard curve.
9. Download data and present as graphs or tables.

4. Notes

1. Whenever possible, we used the TRIzol® Reagent (Invitrogen) for isolation of RNA because of its convenience, good RNA quality and speed. Some plant tissues may not be amenable to isolation of RNA by this method. Other methods for isolation of RNA may be used; however, avoid RNA purification methods that use RNA-binding glass fibre filters that do not recover small RNA species quantitatively (e.g. Qiagen RNeasy® mini and midi kits). If unfamiliar with the method for isolation of RNA, subsequent isolation, quantification and polyacrylamide gel electrophoresis of the low molecular weight RNA fraction can be used to evaluate its quantity and quality.
2. Spectrophotometry followed by gel electrophoresis is still the most widely used method for assessing the RNA yield, purity and quality. Fluorometry (e.g. RiboGreen®, Molecular Probes) can also be used to determine yield, and microfluidic systems (such as Agilent's Bioanalyzer chips) can be used to determine yield and quality.
3. The presented miRNA hydrolysis probe assay utilizes UPL probe #21 (Roche Diagnostics), a universal hydrolysis probe of eight nucleotides including one locked nucleic acid (LNA), that is specific to the stem-loop RT primer backbone. Individual miRNA-specific TaqMan probes may provide increased specificity for quantification of highly similar miRNAs. MicroRNA assay kits for detection of individual miRNAs using individual miRNA-specific TaqMan probes are available commercially (Applied Biosystems). They are designed for quantification of published miRNA sequences available in miRBase (http://www.mirbase.org/).
4. It has been suggested that denaturation of RNA may reduce the yield of cDNA for some miRNAs. In our hands, both non-denatured RNA and RNA denatured by incubation at 65°C for 5 min produced similar results.

5. In standard RT-PCR, endogenous controls are amplified from the same RT template as the genes of interest, thus taking into account the possible variation in efficiency of reverse transcription between samples. In stem-loop RT-PCRs each RT reaction is performed using a miRNA-specific stem-loop RT primer. Therefore, the endogenous controls are transcribed separately.
6. Endogenous control genes are chosen based on expression that is abundant and relatively constant across tissues and cell types. In addition, a suitable control for normalization of miRNA expression would be similar to miRNAs in terms of size and stability and would be amenable to the miRNA assay design. Some classes of small non-coding RNAs (ncRNAs) other than miRNAs are often expressed abundantly and in a stable manner. Several human and mouse snRNAs and snoRNAs were tested across the range of tissues and experimental conditions and confirmed as suitable endogenous controls for quantification of miRNA expression levels (Applied Biosystems). A similar large-scale analysis of plant ncRNA is required to evaluate their suitability for plant miRNA quantification. Currently, chosen endogenous controls need to be validated for a particular experimental design and may include a specific miRNA that demonstrates the least variability across tissues or experimental conditions. In general, evaluation of endogenous controls involves demonstration of relatively abundant and relatively constant expression levels across the tissues and environmental conditions, compared with the RNA input and expression of other housekeeping genes.
7. Melting curve analysis needs to be combined with gel electrophoresis. Due to the small size of the fragment, a primer dimer product generated from the 'minus RT' and 'no template' controls often has a very similar T_m to that of the appropriate miRNA amplification fragment. This becomes an issue with lowly abundant miRNAs that require a large number of PCR amplification cycles. In that case, the UPL assay is recommended.
8. Alternatively, use a cDNA sample with the highest level of target expression and prepare serial dilutions.
9. This method assumes that the efficiency of target and reference gene amplification is identical and equal to 2 (the amount of PCR product doubles during each cycle). In reality, the efficiency is often lower because of a number of different factors. Efficiency correction is required for more reliable data. The software calculates the efficiency from the slope of the standard curve.

References

1. Hannon, G. J. (2002) RNA interference. *Nature* **418**, 244–251.
2. Bartel, D. P. (2004) MicroRNAs: genomics, biogenesis, mechanism, and function. *Cell* **116**, 281–297.
3. Bartel, B. and Bartel, D. P. (2003) MicroRNAs: at the root of plant development? *Plant Physiol.* **132**, 709–717.
4. Mallory, A. C. and Vaucheret, H. (2006) Functions of microRNAs and related small RNAs in plants. *Nat. Genet.* **38**(Suppl), S31–S36.
5. Voinnet, O. (2009) Origin, biogenesis, and activity of plant microRNAs. *Cell* **136**, 669–687.
6. Aukerman, M. J. and Sakai, H. (2003) Regulation of flowering time and floral organ identity by a MicroRNA and its *APETALA2*-like target genes. *Plant Cell* **15**, 2730–2741.
7. Chen, X. (2004) A microRNA as a translational repressor of *APETALA2* in *Arabidopsis* flower development. *Science* **303**, 2022–2025.
8. Llave, C., Xie, Z., Kasschau, K. D., and Carrington, J. C. (2002) Cleavage of *Scarecrow-like* mRNA targets directed by a class of *Arabidopsis* miRNA. *Science* **297**, 2053–2056.
9. Palatnik, J. F., Allen, E., Wu, X., Schommer, C., Schwab, R., Carrington, J. C., and Weigel, D. (2003) Control of leaf morphogenesis by microRNAs. *Nature* **425**, 257–263.
10. Lanet, E., Delannoy, E., Sormani, R., Floris, M., Brodersen, P., Crete, P., Voinnet, O., and Robaglia, C. (2009) Biochemical evidence for translational repression by *Arabidopsis* microRNAs. *Plant Cell* **21**, 1762–1768.
11. Bao, N., Lye, K. W., and Barton, M. K. (2004) MicroRNA binding sites in *Arabidopsis* class III HD-ZIP mRNAs are required for methylation of the template chromosome. *Dev. Cell* **7**, 653–662.
12. Llave, C., Kasschau, K. D., Rector, M. A., and Carrington, J. C. (2002) Endogenous and silencing-associated small RNAs in plants. *Plant Cell* **14**, 1605–1619.
13. Rhoades, M. W., Reinhart, B. J., Lim, L. P., Burge, C. B., Bartel, B., and Bartel, D. P. (2002) Prediction of plant microRNA targets. *Cell* **110**, 513–520.
14. Sunkar, R. and Zhu, J. K. (2004) Novel and stress-regulated microRNAs and other small RNAs from *Arabidopsis. Plant Cell* **16**, 2001–2019.
15. Sunkar, R., Kapoor, A., and Zhu, J. K. (2006) Posttranscriptional induction of two Cu/Zn superoxide dismutase genes in *Arabidopsis* is mediated by downregulation of miR398 and important for oxidative stress tolerance. *Plant Cell* **18**, 2051–2065.
16. Juarez, M. T., Kui, J. S., Thomas, J., Heller, B. A., and Timmermans, M. C. (2004) microRNA-mediated repression of *rolled leaf1* specifies maize leaf polarity. *Nature* **428**, 84–88.
17. Kidner, C. A. and Martienssen, R. A. (2004) Spatially restricted microRNA directs leaf polarity through ARGONAUTE1. *Nature* **428**, 81–84.
18. Tang, G., Reinhart, B. J., Bartel, D. P., and Zamore, P. D. (2003) A biochemical framework for RNA silencing in plants. *Genes Dev.* **17**, 49–63.
19. Mallory, A. C., Reinhart, B. J., Jones-Rhoades, M. W., Tang, G., Zamore, P. D., Barton, M. K., and Bartel, D. P. (2004) MicroRNA control of *PHABULOSA* in leaf development: importance of pairing to the microRNA 5′ region. *EMBO J.* **23**, 3356–3364.
20. Yoo, B. C., Kragler, F., Varkonyi-Gasic, E., Haywood, V., Archer-Evans, S., Lee, Y. M., Lough, T. J., and Lucas, W. J. (2004) A systemic small RNA signaling system in plants. *Plant Cell* **16**, 1979–2000.
21. Pant, B. D., Buhtz, A., Kehr, J., and Scheible, W. R. (2008) MicroRNA399 is a long-distance signal for the regulation of plant phosphate homeostasis. *Plant J.* **53**, 731–738.
22. Parizotto, E. A., Dunoyer, P., Rahm, N., Himber, C., and Voinnet, O. (2004) In vivo investigation of the transcription, processing, endonucleolytic activity, and functional relevance of the spatial distribution of a plant miRNA. *Genes Dev.* **18**, 2237–2242.
23. Alvarez, J. P., Pekker, I., Goldshmidt, A., Blum, E., Amsellem, Z., and Eshed, Y. (2006) Endogenous and synthetic microRNAs stimulate simultaneous, efficient, and localized regulation of multiple targets in diverse species. *Plant Cell* **18**, 1134–1151.
24. Lu, C., Tej, S. S., Luo, S., Haudenschild, C. D., Meyers, B. C., and Green, P. J. (2005) Elucidation of the small RNA component of the transcriptome. *Science* **309**, 1567–1569.
25. Chen, C., Ridzon, D. A., Broomer, A. J., Zhou, Z., Lee, D. H., Nguyen, J. T., Barbisin, M., Xu, N. L., Mahuvakar, V. R., Andersen, M. R., Lao, K. Q., Livak, K. J., and Guegler, K. J. (2005) Real-time

quantification of microRNAs by stem-loop RT-PCR. *Nucleic Acids Res.* **33**, e179.

26. Tang, F., Hajkova, P., Barton, S. C., Lao, K., and Surani, M. A. (2006) MicroRNA expression profiling of single whole embryonic stem cells. *Nucleic Acids Res.* **34**, e9.

27. Varkonyi-Gasic, E., Wu, R., Wood, M., Walton, E. F., and Hellens, R. P. (2007) Protocol: a highly sensitive RT-PCR method for detection and quantification of microRNAs. *Plant Methods* **3**, 12.

Chapter 11

Large-Scale Sequencing of Plant Small RNAs

William P. Donovan, Yuanji Zhang, and Miya D. Howell

Abstract

Deep sequencing technologies have become very powerful tools in the identification and quantification of small RNAs involved in gene regulation. Small interfering RNA (siRNA) and miRNA are two classes of DCL-dependent small RNAs known to affect phenotype, developmental regulation, and various traits in plants. These small RNAs function by selectively repressing gene expression mainly by guiding cleavage, resulting in degradation of target transcripts. In this chapter, we describe a method for preparation of 5′-phosphate-dependent small RNA libraries, a hallmark of RNase III-like DCL products, for high-throughput sequencing, and recommendations for small RNA analysis. This method is useful for determining small RNA involvement in critical pathways in plants, identifying and quantifying novel small RNAs, and examining small RNA global expression patterns.

Key words: DNA sequencing, high-throughput sequencing, Illumina, miRNA, siRNA, small RNA library construction, small RNA reads.

1. Introduction

The plant small RNA repertoire is largely dominated by short interfering RNAs (siRNAs) and miRNAs approximately 21–24 nucleotides in length. Both siRNAs and miRNAs are important riboregulators of gene expression in plants (1–3). siRNAs function in transcriptional gene silencing mechanisms, resulting in chromatin modifications and DNA methylation (4, 5), while miRNAs and some siRNA classes (6–8) are post-transcriptional gene regulators that generally function by triggering cleavage of target transcripts. miRNAs arise from imperfect self-complementary foldback structures which are subsequently cleaved in a DCL1-dependent manner to produce a double-stranded duplex with two nucleotide 3′ overhangs (9). The miRNA strand is loaded into an

H. Kodama, A. Komamine (eds.), *RNAi and Plant Gene Function Analysis*, Methods in Molecular Biology 744,
DOI 10.1007/978-1-61779-123-9_11, © Springer Science+Business Media, LLC 2011

RNA-induced silencing complex (RISC) and functions by either guiding cleavage of a target transcript or inhibiting translation (10, 11).

Small RNA-directed gene silencing regulates many cellular processes including hormone signaling, development, stress responses, and genome maintenance (3, 12–14). High-throughput sequencing technologies have greatly improved our effort in the identification and characterization of plant-derived small RNAs from crops and are particularly useful in identifying small RNAs involved in various traits (i.e., yield, biomass, and insect tolerance), developmental responses, and biotic and abiotic stress responses. Analysis of small RNA data from these large datasets is critical to further our understanding of gene regulation of small RNAs.

In this chapter, we describe the methods for 5′ phosphate-dependent small RNA library construction for high-throughput sequencing, and the workflow process and requirements for small RNA data analysis. Data from these small RNA libraries are useful for determining small RNA involvement in critical pathways in crop plants. Emphasis will be placed on system requirements for small RNA analysis, and methods to screen the libraries to filter out non-essential sequence reads. Upstream instrumentation-specific analysis such as base-calling and detailed downstream project-specific analysis such as novel siRNA/miRNA discovery will not be covered in this chapter.

2. Materials

2.1. Small RNA Isolation

1. Harvest plant tissue using liquid nitrogen and forceps.
2. RNA extraction: Trizol® reagent (Invitrogen) (store at 4°C), chloroform, isopropanol, ethanol, DEPC-treated water, 15 mL/50 mL polypropylene RNase-/DNase-free centrifuge tubes, non-stick RNase-free microfuge tubes.
3. Small RNA isolation and purification: Novex® TBE–urea sample buffer (2×) (Invitrogen) (store at 4°C), 17% TBE–urea gel, 1× TBE running buffer, XCell SureLock® Mini-Cell CE mark (Invitrogen), 18-nt RNA size standard (5′-rGrUrArCrGrCrGrGrGrUrUrUrArArArCrGrA-3′) (Integrated DNA Technologies) (store master stock at –80°C), 26-nt RNA size standard (5′-rGrUrArCrGrCrGrGrGrUrUr UrArArArCrGrArUrGrCrArArCrGrU-3′) (Integrated DNA Technologies) (store master stock at –80°C), GelStar Nucleic Acid Stain (Lonza), Blue light box (i.e., Invitrogen Safe Imager™ 2.0 Blue-Light Transilluminator), Performa® DTR gel filtration cartridges (Edge BioSystems) (store at

4°C), disposable plastic pestles (i.e., Kimble Chase KONTES RNase-free pestles VWR), GlycoBlue™ (Ambion), and 3 M sodium acetate.

2.2. Small RNA Cloning and Assessment

1. Adaptor ligation/purification: T4 RNA ligase (Ambion), T4 RNA ligase 2, truncated (New England Biolabs), 3′ adaptor (5′-/5rApp/ATCTCGTATGCCGTCTTCTGCTTG/3ddC/-3′) (*see* **Note 1**), 5′ RNA adaptor (5′-rGrUrUrCrArGrArGrUrUrCrUrArCrArGrUrCrCrGrArCrGrArUrC-3′) (*see* **Note 1**).
2. Reverse transcription: SuperScript® III Reverse Transcriptase (Invitrogen), RT primer (5′-CAAGCAGAAGACGGCATACGA-3′) (*see* **Note 1**).
3. PCR amplification and purification: Phusion™ High-Fidelity PCR Kit (New England Biolabs), *Pme*I, DNA loading dye, 10% acrylamide/TBE gel, Low Molecular Weight DNA Ladder (New England Biolabs), reverse primer (5′-CAAGCAGAAGACGGCATACGA-3′) (*see* **Note 1**), forward primer (5′-AATGATACGGCGACCACCGACAGGTTCAGAGTTCTACAGTCCGA-3′) (*see* **Note 1**).
4. Quality assessment: Zero Blunt® TOPO® PCR Cloning Kit (Invitrogen).

3. Methods

3.1. Total RNA Extraction/Purification from Plant Tissue

1. Harvest fresh plant tissue immediately in a tube containing liquid nitrogen (*see* **Note 2**).
2. Grind tissue to fine powder using mortar and pestle in liquid nitrogen.
3. Add 10 mL of Trizol reagent per 1 g tissue to frozen powder in mortar. Grind/mix until homogenous. Allow suspension to thaw and transfer to a fresh polypropylene tube (*see* **Note 3**).
4. Add 3 mL of chloroform to 10 mL of tissue:Trizol suspension. Mix well by vortex or shaking vigorously. Incubate tubes at room temperature for 3 min.
5. Centrifuge at 10,000×*g* for 10 min at 4°C. Transfer the top aqueous phase to a new polypropylene tube.
6. Repeat steps 4–5 until the interface between the aqueous and organic layers is no longer present (*see* **Note 4**).
7. Add 1:1 volume of cold isopropanol to the top aqueous layer in a fresh tube. Incubate samples for 30 min at room temperature. Centrifuge at 10,000×*g* for 10 min at 4°C.

8. Note the presence of pellet. Remove the supernatant from the tube. Wash the pellet with 80% cold ethanol. Briefly allow pellet to air-dry.
9. Resuspend the pellet in a minimal volume of DEPC-treated water. Transfer RNA to RNase-free, non-stick, 1.5-mL microtube.
10. Determine the concentration of total RNA using A260/280 (*see* **Note 5**). Store RNA at –80°C.

3.2. Small RNA Isolation

1. Mix approximately 100–500 μg of total RNA (in a volume of 50–100 μL DEPC-treated water) with an equal volume of Novex® TBE–urea sample buffer (2×).
2. Heat sample at 70°C for 5 min and place on ice for 1 min.
3. Prepare a 17% polyacrylamide gel containing 7 M urea in 0.5× TBE and pre-run at 200 V for 20 min in 1× TBE buffer. Flush wells several times with 1× TBE buffer to remove urea prior to loading samples.
4. Load sample into wells. In nearby adjacent lanes, load 20 pmol each of 18 and 26 nt RNA size standards (*see* **Note 6**).
5. Electrophorese at 200 V in 1× TBE for 1 h or until the bromophenol blue dye is near the bottom of the gel.
6. Remove the gel from the apparatus and transfer to an RNase-free container containing 1× TBE and 1× GelStar. Gently shake the gel in solution for 10 min at room temperature. Rinse gel twice in 1× TBE.
7. Image and slice gel with a scalpel containing the 18–26-nt small RNA region (use RNA size standards to determine migration) using a Blue-Light Transilluminator box.
8. Transfer the gel slice (two-well width max., approximately 200 mg) to a non-stick, 1.5-mL microtube.
9. Isolate RNA using a DTR gel filtration cartridge. Add 350 μL of DEPC-treated H_2O to the gel slice.
10. Crush the gel to a fine slurry with a disposable plastic pestle. Incubate the slurry at 70°C for 10 min. Vortex the suspension for approximately 20 s.
11. Prepare the DTR gel filtration cartridge. Centrifuge the cartridge in a swinging bucket microfuge (i.e., Eppendorf #5471C) at 850×*g* for 3 min. Transfer the cartridge to a non-stick 1.5-mL microtube.
12. Pour the slurry into the DTR cartridge. Centrifuge at 850×*g* for 3 min.
13. Precipitate the RNA. Add 2.3 μL of 15 mg/mL GlycoBlue (100 μg/mL), 1/10 volume of 3 M sodium acetate (pH 5.2), and 2.5× volume of 100% ethanol. Vortex briefly.

14. Freeze at –80°C for 2 h or –20°C overnight. Centrifuge at 20,000×*g* for 30 min at 4°C.
15. Remove the supernatant. Note the position of the blue pellet. Wash with 80% ethanol. Spin for 2 min.
16. Remove the ethanol and air-dry the pellet. Resuspend (and combine pellets if necessary) in a total volume of 16.5 μL DEPC-treated water.
17. Proceed to 3′ adaptor ligation.

3.3. 3′ Adaptor Ligation and Purification

1. This method is designed to use specific adaptors from Illumina® (*see* **Note 1**). The 3′ adaptor includes a pre-adenylated 5′-end for ligation using T4 Rnl2-trunc ligase (-ATP). The use of this modified adaptor and truncated T4 RNA ligase 2, which is unable to adenylate the 5′-end of a substrate, prevents ligation of a phosphorylated 5′-end to the 3′-end of RNA. The 3′ adaptor also consists of a 3′-dideoxycytidine (ddC) to prevent self-ligation of the adaptor.
2. The following is added to 16.5 μL of the purified small RNA sample: 1 μL of 100 μM 3′ adaptor, 2 μL of 10× T4 Rnl2 trunc reaction buffer, and 0.5 μL T4 Rnl2 trunc ligase (100 U).
3. Incubate the sample at room temperature for 2 h.
4. Add an equal volume Novex® TBE–urea sample buffer (2×) to sample.
5. Heat the sample at 70°C for 5 min and place on ice for 1 min.
6. Prepare a 17% polyacrylamide gel containing 7 M urea in 0.5× TBE and pre-run at 200 V for 20 min in 1× TBE buffer. Flush wells several times with 1× TBE buffer to remove urea prior to loading samples.
7. Load sample into wells. In nearby adjacent lanes, load 2 pmol each of 37-nt and 42-nt RNA size standards (*see* **Note 6**).
8. Electrophorese at 200 V in 1× TBE for 1 h or until the bromophenol blue dye is near the bottom of the gel.
9. Remove the gel from the apparatus and transfer to an RNase-free container containing 1× TBE and 1× GelStar. Gently shake in solution for 10 min at room temperature. Rinse the gel twice in 1× TBE.
10. Image and slice the gel with scalpel containing the 37–42-nt small RNA region (use RNA size standards to determine migration and cut approximately 2 mm above and below 42 and 37 marker standards) using a Blue-Light Transilluminator box.

11. Transfer the gel slice (two-well width max., approximately 200 mg) to a non-stick 1.5-mL microtube.
12. Isolate RNA using a DTR gel filtration cartridge. Add 350 μL DEPC-treated H_2O to the gel slice.
13. Crush the gel to a fine slurry with a disposable plastic pestle. Incubate the slurry at 70°C for 10 min. Vortex the suspension for approximately 20 s.
14. Prepare the DTR gel filtration cartridge. Centrifuge the cartridge in a swinging bucket microfuge at 850×*g* for 3 min. Transfer the cartridge to a non-stick 1.5-mL microtube.
15. Precipitate the RNA. Add 2.3 μL of 15 mg/mL GlycoBlue (100 μg/mL), 1/10 volume 3 M sodium acetate (pH 5.2), and 2.5× volume 100% ethanol. Vortex briefly.
16. Freeze at –80°C for 2 h or –20°C overnight. Centrifuge at 20,000×*g* for 30 min at 4°C.
17. Remove the supernatant. Note the position of the blue pellet. Wash with 80% ethanol. Spin for 2 min.
18. Remove the ethanol and air-dry the pellet. Resuspend (and combine pellets if necessary) in total volume of 16.5 μL DEPC-treated water.
19. Proceed to 5′ adaptor ligation.

3.4. 5′ Adaptor Ligation and Purification

1. To 16.5 μL purified sRNA-3′ adaptor product, add 2 μL T4 RNA ligation buffer with ATP, 1 μL of 100 μM 5′ adaptor (100 pmol), and 0.5 μL T4 RNA ligase (2.5 U).
2. Incubate the sample at 37°C for 1 h.
3. Add 80 μL H_2O.
4. Precipitate the RNA. Add 0.7 μL of 15 mg/mL GlycoBlue (100 μg/mL), 1/10 volume 3 M sodium acetate (pH 5.2), and 2.5× volume 100% ethanol. Vortex briefly.
5. Remove the supernatant and note the position of the blue pellet. Wash with 80% ethanol. Spin for 2 min.
6. Remove the ethanol and air-dry the pellet. Resuspend the pellet in 11 μL DEPC-treated water.
7. Proceed to reverse transcription.

3.5. Reverse Transcription

1. Transfer 11 μL of the ligated product to a 0.2-mL PCR tube.
2. Add 1 μL each of 10 μM RT primer and 10 mM dNTP mix.
3. Incubate the tube at 65°C for 5 min. Transfer to ice for 1 min.
4. Proceed with reverse transcription in a 20-μL reaction using Superscript III RT according to manufacturer's instructions.

5. After the reaction is complete, incubate at 70°C for 5 min to inactivate. Proceed to PCR amplification.

3.6. PCR Amplification and Purification

1. Add 17 μL water, 10 μL of 5× Phusion™ HF buffer, 1 μL each of 10 μM forward and reverse primers, 1 μL of 10 mM dNTP mix, and 0.5 μL Phusion HF DNA polymerase (1 U) to 20 μL product from reverse transcription (*see* **Note** 7).
2. PCR is performed at 98°C for 1 min, then 15 cycles of 98°C for 10 s, 60°C for 30 s, and 72°C for 15 s, with a final extension at 72°C for 10 min.
3. (Steps 3–6 optional. *See* **Note 6**.) Add 50 μL H_2O to reaction mix and precipitate the DNA. Add 0.7 μL of 15 mg/mL GlycoBlue (100 μg/mL), 1/10 volume 3 M sodium acetate (pH 5.2), and 2× volume 100% ethanol. Vortex briefly.
4. Remove the supernatant and note the position of the blue pellet. Wash with 80% ethanol. Spin for 2 min.
5. Remove the ethanol and air-dry the pellet. Resuspend the pellet in 17.5 μL of H_2O.
6. Digest the sample with *Pme*I to remove any incorporated RNA size standards from sample.
7. Add DNA loading buffer to sample. Load sample onto 10% acrylamide/TBE gel. Load Low Molecular Weight DNA Ladder in adjacent lanes.
8. Electrophorese at 200 V in 1× TBE for 45 min or until the bromophenol blue dye is near the bottom of the gel.
9. Remove the gel from the apparatus and transfer to an RNase-free container containing 1× TBE and 1× GelStar. Gently shake in solution for 10 min at room temperature. Rinse the gel twice in 1× TBE.
10. Image and slice the gel with scalpel containing 90-bp PCR product using a Blue-Light Transilluminator box.
11. Transfer the gel slice (two-well width max., approximately 200 mg) to a non-stick 1.5-mL microtube.
12. Isolate PCR product using a DTR gel filtration cartridge. Add 350 μL DEPC-treated H_2O to the gel slice.
13. Crush the gel to a fine slurry with a disposable plastic pestle. Incubate the slurry at 70°C for 10 min. Vortex the suspension for approximately 20 s.
14. Prepare the DTR gel filtration cartridge. Centrifuge the cartridge in a swinging bucket microfuge (i.e., Eppendorf #5471C) at 850×*g* for 3 min. Transfer the cartridge to a non-stick 1.5-mL microtube.

15. Precipitate the DNA. Add 2.3 μL of 15 mg/mL GlycoBlue (100 μg/mL), 1/10 volume 3 M sodium acetate (pH 5.2), and 2.5× volume 100% ethanol. Vortex briefly.
16. Freeze at –80°C for 2 h or –20°C overnight. Centrifuge at 20,000×*g* for 30 min at 4°C.
17. Remove the supernatant and note the position of the blue pellet. Wash with 80% ethanol. Spin for 2 min.
18. Remove the ethanol and air-dry the pellet. Resuspend (and combine pellets if necessary) in a total volume of 12 μL of 10 mM Tris–HCl (pH 7.4).
19. Estimate the concentration of the purified PCR product. Load 0.5 μL of product on 10% acrylamide/TBE gel using a DNA ladder with mass standards. The concentration of the final product should estimate between 5 and 10 ng/μL.

3.7. Assessment of Quality

1. If possible, load sample on Agilent Technologies Bioanalyzer. Check the library by conventional cloning and dideoxy sequencing prior to high-throughput sequencing.
2. TOPO clone approximately 2 ng of the final PCR product into pCR Blunt II-TOPO according to manufacturer's instructions.
3. Sequence using conventional technologies. A good library will consist of samples containing both the 5′ and 3′ adaptor sequences, unique inserts between 18 and 26 bp, and no self-ligating clones. **Figure 11.1** represents a schematic of the small RNA library procedure.

3.8. High-Throughput Sequencing

The constructed small RNA libraries are subjected to high-throughput sequencing for small RNA analysis. Traditionally conventional cloning into plasmid vectors was used to sequence small RNA reads, but with the advancement of sequencing technologies, small RNAs can be directly amplified and sequenced resulting in a much higher throughput. Several high-throughput sequencing technologies have been used to sequence small RNA reads including MPSS, 454, Illumina, and SOLiD. Here we focus on the single-read sequencing assay from Illumina®, which utilizes a four-color DNA sequencing by synthesis (SBS) approach. Sequence information from libraries prepared using any of above approaches provides a rich resource for small RNA reads that can subsequently be analyzed for expression patterns (*see* **Note 8**).

3.9. System Requirements and Workflow for Small RNA Analysis

Because of the large scale of raw data retrieved from current high-throughput sequencing platforms, it is recommended that data analysis be performed using the Unix computer operating system to utilize the available command line utilities. Other tools essential for analysis include a high-level programming language

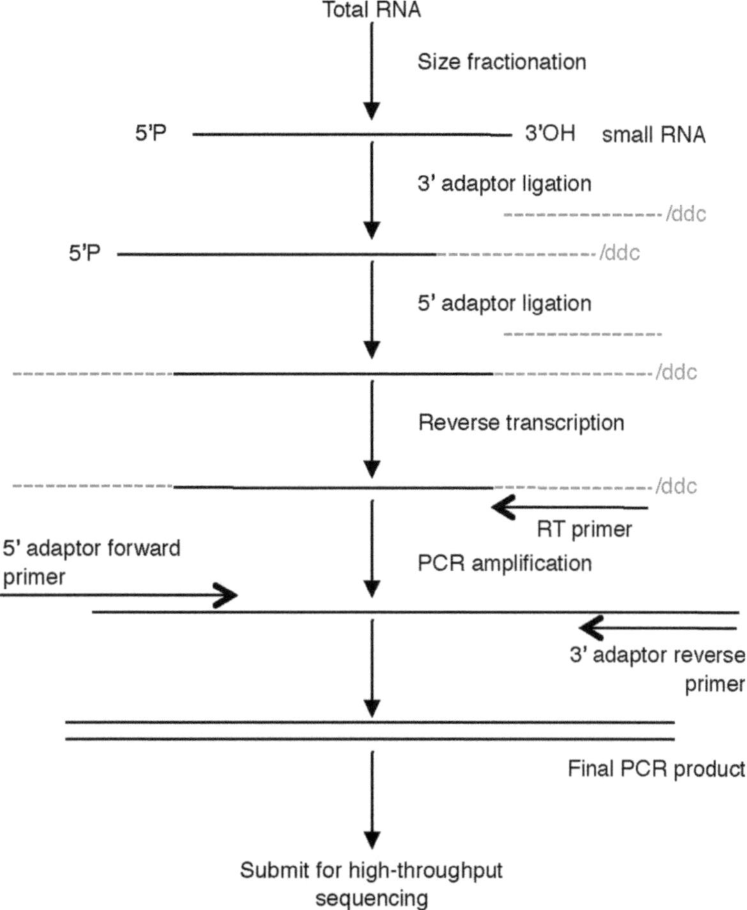

Fig. 11.1. Schematic of the small RNA library preparation method. Total RNA is extracted from plant tissue followed by small RNA purification by polyacrylamide gel-based fractionation. Adaptors are ligated to the 3′- and 5′-ends of the small RNA. The subsequent product undergoes reverse transcription and the cDNA is amplified. Small RNA libraries are then subjected to quality control checks prior to submission for high-throughput sequencing.

such as Perl, a relational database system to manage data, and an http server for Web-based tools to allow users to access and mine the data. A software bundle such as LAMP (Linux, Apache, MySQL, Perl) is an ideal solution stack. Perl is the de facto standard choice in programming languages and is also popular in CGI programming for interfacing between servers and application software.

Raw sequence data from the sequencing run should be made available in FASTA format after base calling and quality score filtering using software provided or suggested by the sequencing platform vendor. A multiple-step analysis pipeline is recommended to further evaluate sequence quality, parse small RNA reads from the raw sequence data, and classify and characterize

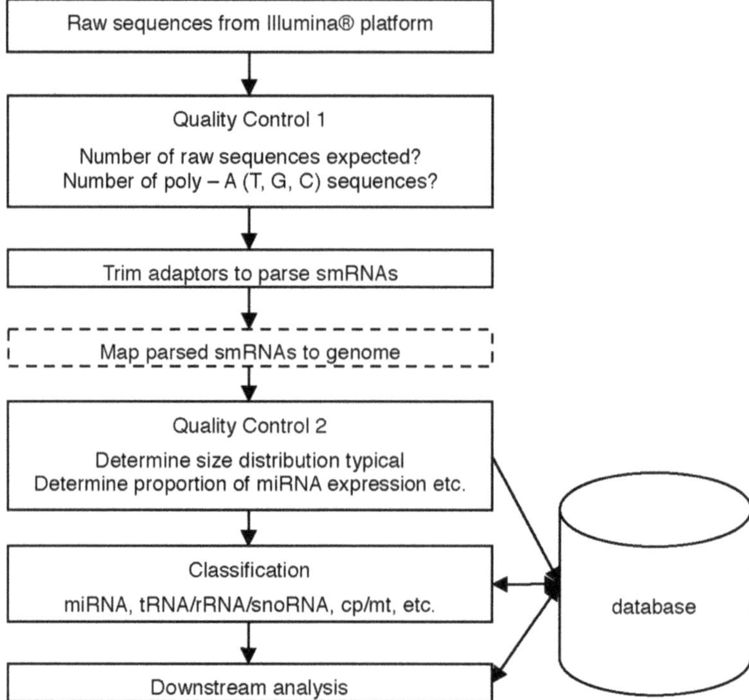

Fig. 11.2. Workflow for small RNA analysis. In this example, the analysis pipeline is not restricted to any particular sequencing platform. Upon retrieval of raw sequence information, initial quality control assesses the number of total raw reads, the number of reads containing ambiguous base calls, and the number of reads containing poly-N fragments. Small RNA reads between 18 and 26 nt in length are parsed after identification of the 3′ adaptor sequence and trimmed from raw sequences containing 5′ and 3′ adaptor sequence. Small RNA reads are mapped to a genome reference sequence if available. The size distribution and expression levels of the remaining small RNA reads are determined, and any small RNA reads corresponding to known miRNA sequences are determined. These small RNAs are then loaded into a defined database to support further analysis including classification and further annotation of small RNAs, genome mapping, and novel small RNA identification.

small RNA reads. A database should be designed to manage the data, assist the analysis, and facilitate data mining and data accessibility. **Figure 11.2** illustrates the workflow of the recommended bioinformatics pipeline:

1. Because of the length of time and resources required to prepare samples, sequence libraries, and analyze data, the first task during data analysis should involve determining whether systematic errors are present in the sequence population. The total number of raw sequences should match the throughput of the sequencing platform. For example, the output from one lane using the current Illumina® Genome Analyzer platform is approximately 1×10^7 reads.

2. Small RNA reads ranging between 18 and 26 nucleotides in length are parsed and trimmed from raw sequences by

identification of the 5′ and 3′ adaptor sequences (*see* **Note 9**). Raw reads of poly-N nucleotide strings should account for < 5% of overall raw sequences.

3. If the genome or the reference sequence is available for the source organism, map the small RNA reads to the reference sequence to filter any contaminated or background sequence reads. Most unique small RNA reads (>80%) should map perfectly to the reference sequence (*see* **Note 10**).
4. Measure the size distribution of both unique and total abundance small RNA reads mapping to the reference sequence. Small RNA reads should have a characteristic bimodal size distribution, i.e., two peaks at 21 and 24 nucleotides (nt) representing miRNA and siRNA classes. For some species/tissues, small RNA peaks may also be apparent at 22 nt (**Fig. 11.3**).

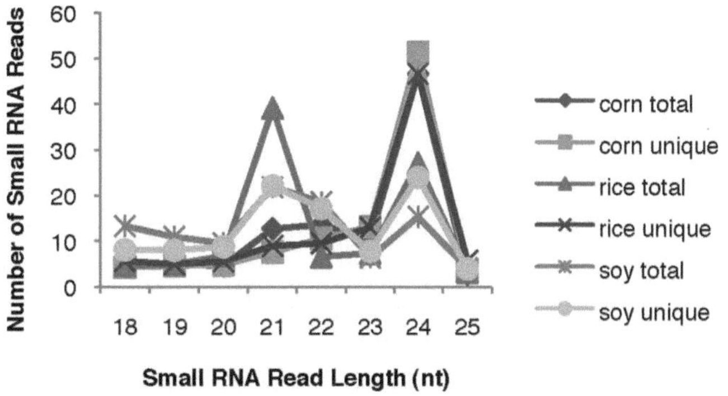

Fig. 11.3. Representative size distribution of small RNA reads from small RNA libraries prepared from rice, corn, or soy seedling tissue. The numbers of total and unique (repeat normalized to account for redundancy) small RNAs from individual libraries ranging from 18 to 25 nt in length are reported.

5. Measure expression levels of small RNA reads corresponding to known miRNA sequences. Compare the parsed small RNA reads to a miRNA dataset (for example, miRBase at http://microrna.sanger.ac.uk/, the public repository for annotated miRNAs from plants, animals, and viruses) (15) (*see* **Note 11**).
6. Load small RNA reads into a defined database.
7. Further analysis of the small RNA reads includes classification and further annotation of small RNAs, genome mapping, and novel small RNA identification. Variations in library sequence data could be expected in libraries prepared from transgenic or viral-infected plant tissue as compared to nontransgenic library preparations (*see* **Note 12**).

Sequencing technologies are ever-evolving and will continue to produce a greater throughput of data. A database capturing essential sequence information is the best approach for enabling data analysis of large-scale sequencing projects.

4. Notes

1. Oligonucleotide sequences and adaptors © 2007–2009 Illumina, Inc. All rights reserved. Illumina does not guarantee results or performance if oligonucleotides are purchased elsewhere.
2. All required materials should be RNase free, and materials should be autoclaved prior to use. Tissue must be frozen immediately and remain frozen until Trizol addition to avoid RNA degradation and loss of yield.
3. This protocol describes the use of Trizol, an organic extraction reagent. Trizol may be added directly to the frozen mortar and pestle and ground into the frozen tissue, or frozen ground tissue may be added to a conical containing premeasured Trizol solution. In some instances, Trizol is not the recommended reagent for RNA extraction. For example, cotton RNA extraction is difficult due to the high levels of secondary metabolites. If RNA extraction using Trizol reagent results in low yield, other methods such as the hot borate method have been successfully used (16).
4. Generally three or more chloroform steps are performed. Some plant tissues yield discolored RNA pellets resulting from polysaccharides or secondary metabolites. This can be avoided by using more Trizol and performing additional chloroform steps. High-quality RNA pellets appear white.
5. OD *A*260/280 for high-quality RNA will be 2.0 or slightly higher.
6. The RNA size standards for small RNA isolation are nonspecific unmodified ribonucleotide sequences. These sequences contain a *Pme*I site (GTTT^AAAC) for removal of standard marker contamination. As long as size standards are not loaded in adjacent lanes as sample wells, digestion of the final PCR product with *Pme*I is not necessary. It is not recommended to load RNA from more than two different library preparations on a single gel.
7. PCR amplification of the reverse-transcribed cDNA allows for selective enrichment of cDNA fragments with both 5′ and 3′ adaptor molecules. The PCR is performed with primers that overlap the Illumina® adaptors and kept at a

low number of cycles to prevent enrichment of nonspecific products.

8. In-house library data using both 454 and Illumina technologies suggest there is a 3′ adaptor ligation bias to selected nucleotides at the 3′-end of the small RNA. This was also observed by Linsen et al. (17). Therefore normalization to control libraries (i.e., using synthetic oligonucleotide standards (18)) and determining relative abundance levels will be more useful than absolute small RNA numbers.
9. The purpose of parsing small RNA reads from raw reads is to separate the adaptor sequence information from the actual plant-derived small RNA sequence. For example, if using the Illumina® platform, the 3′ adaptor sequence is identified in the raw sequence and used to determine the small RNA sequence. Parsing includes identifying the 3′ adaptor and eliminating sequence reads, resulting in direct ligation of 5′ adaptor–3′ adaptor or 3′ adaptor–3′ adaptor. Small RNA reads containing ambiguous letters (i.e., N) or that have a stretch of greater than 10 nucleotides of the same base should be excluded from the analysis.
10. For species whose genome sequence is not available, other types of sequences such as ESTs can be used, and percentage of unique smRNAs mapping to the sequences should be proportional to the genome coverage of ESTs or other sequences used for the mapping.
11. Errors or quality problems in the sequence data are likely to have occurred when the following observations are made: a large proportion of animal miRNAs match to a plant-specific small RNA library; an unexpected low number of reads map to known miRNAs; reads mapping to known miRNA families for the plant/tissue type are inconsistent with published results for miRNA expression.
12. Structural smRNAs originating from tRNA/rRNA/snRNA/snoRNA are generally believed to be nonfunctional and it is a common practice to filter them out. However, for many species the annotation of these structural RNAs is either incomplete or not available at all. Recent studies indicate that snoRNAs may produce siRNAs which may function like miRNAs (19, 20). In the filamentous fungus *Neurospora crassa*, a novel class of siRNAs named "qiRNA" is derived primarily from ribosomal DNA locus and may play a role in DNA damage response (21). Highly abundant Dicer-dependent small RNAs have also been identified from HeLa-specific tRNA

fragments (22). Rather than filtering out these structural smRNAs, a conservative alternative is to flag them in the database to create flexibility in downstream analyses.

Acknowledgments

We would like to thank Sara Heisel, Hong Liu, Mingya Huang, Zijin Du, Phil Latreille, and Xuefeng Zhou for their contributions to our studies involving high-throughput sequencing. We would also like to thank Mia Unson, Julia Gloekner, and Greg Heck for their helpful comments on this chapter. This work was supported by Monsanto Company.

References

1. Voinnet, O. (2009) Origin, biogenesis, and activity of plant microRNAs. *Cell* **136**, 669–687.
2. Vaucheret, H. (2006) Post-transcriptional small RNA pathways in plants: Mechanisms and regulations. *Genes Dev.* **20**, 759–771.
3. Mallory, A. C. and Vaucheret, H. (2006) Functions of microRNAs and related small RNAs in plants. *Nat. Genet.* **38**, S31–S36.
4. Daxinger, L., Kanno, T., Bucher, E., van der Winden, J., Naumann, U., Matzke, A. J., and Matzke, M. (2009) A stepwise pathway for biogenesis of 24-nt secondary siRNAs and spreading of DNA methylation. *EMBO J.* **28**, 48–57.
5. Matzke, M., Kanno, T., Huettel, B., Daxinger, L., and Matzke, A. J. (2007) Targets of RNA-directed DNA methylation. *Curr. Opin. Plant Biol.* **10**, 512–519.
6. Allen, E., Xie, Z., Gustafson, A. M., and Carrington, J. C. (2005) microRNA-directed phasing during trans-acting siRNA biogenesis in plants. *Cell* **121**, 207–221.
7. Schwab, R., Maizel, A., Ruiz-Ferrer, V., Garcia, D., Bayer, M., Crespi, M., Voinnet, O., and Martienssen, R. A. (2009) Endogenous TasiRNAs mediate non-cell autonomous effects on gene regulation in *Arabidopsis thaliana*. *PLoS One* **4**, e5980.
8. Vazquez, F., Vaucheret, H., Rajagopalan, R., Lepers, C., Gasciolli, V., Mallory, A. C., Hilbert, J. L., Bartel, D. P., and Crete, P. (2004) Endogenous trans-acting siRNAs regulate the accumulation of *Arabidopsis* mRNAs. *Mol. Cell* **16**, 69–79.
9. Xie, Z., Johansen, L. K., Gustafson, A. M., Kasschau, K. D., Lellis, A. D., Zilberman, D., Jacobsen, S. E., and Carrington, J. C. (2004) Genetic and functional diversification of small RNA pathways in plants. *PLoS Biol.* **2**, E104.
10. Brodersen, P., Sakvarelidze-Achard, L., Bruun-Rasmussen, M., Dunoyer, P., Yamamoto, Y. Y., Sieburth, L., and Voinnet, O. (2008) Widespread translational inhibition by plant miRNAs and siRNAs. *Science* **320**, 1185–1190.
11. Lanet, E., Delannoy, E., Sormani, R., Floris, M., Brodersen, P., Crete, P., Voinnet, O., and Robaglia, C. (2009) Biochemical evidence for translational repression by *Arabidopsis* microRNAs. *Plant Cell* **21**, 1762–1768.
12. Moldovan, D., Spriggs, A., Yang, J., Pogson, B. J., Dennis, E. S., and Wilson, I. W. (2009) Hypoxia-responsive microRNAs and trans-acting small interfering RNAs in *Arabidopsis*. *J. Exp. Bot.* **61**, 165–177.
13. Ruiz-Ferrer, V. and Voinnet, O. (2009) Roles of plant small RNAs in biotic stress responses. *Annu. Rev. Plant Biol.* **60**, 485–510.
14. Ben Amor, B., Wirth, S., Merchan, F., Laporte, P., d'Aubenton-Carafa, Y., Hirsch, J., Maizel, A., Mallory, A., Lucas, A., Deragon, J. M., Vaucheret, H., Thermes, C., and Crespi, M. (2009) Novel long non-protein coding RNAs involved in *Arabidopsis* differentiation and stress responses. *Genome Res.* **19**, 57–69.
15. Griffiths-Jones, S., Saini, H. K., van Dongen, S., and Enright, A. J. (2008) miRBase: Tools for microRNA genomics. *Nucl. Acids Res.* **36**, D154–D158.
16. Wan, C. Y. and Wilkins, T. A. (1994) A modified hot borate method significantly enhances

the yield of high-quality RNA from cotton (*Gossypium hirsutum* L.). *Anal. Biochem.* **223**, 7–12.

17. Linsen, S. E., de Wit, E., Janssens, G., Heater, S., Chapman, L., Parkin, R. K., Fritz, B., Wyman, S. K., de Bruijn, E., Voest, E. E., Kuersten, S., Tewari, M., and Cuppen, E. (2009) Limitations and possibilities of small RNA digital gene expression profiling. *Nat. Methods* **6**, 474–476.
18. Fahlgren, N., Sullivan, C. M., Kasschau, K. D., Chapman, E. J., Cumbie, J. S., Montgomery, T. A., Gilbert, S. D., Dasenko, M., Backman, T. W., Givan, S. A., and Carrington, J. C. (2009) Computational and analytical framework for small RNA profiling by high-throughput sequencing. *RNA* **15**, 992–1002.
19. Ender, C., Krek, A., Friedländer, M. R., Beitzinger, M., Weinmann, L., Chen, W., Pfeffer, S., Rajewsky, N., and Meister, G. (2008) A Human snoRNA with microRNA-like functions. *Mol. Cell* **32**, 519–528.
20. Taft, R. J., Glazov, E. A., Lassmann, T., Hayashizaki, Y., Carninci, P., and Mattick, J. S. (2009) Small RNAs derived from snoRNAs. *RNA* **15**, 1233–1240.
21. Lee, H.-C., Chang, S.-S., Choudhary, S., Aalto, A. P., Maiti, M., Bamford, D. H., and Liu, Y. (2009) qiRNA is a new type of small interfering RNA induced by DNA damage. *Nature* **459**, 274–277.
22. Cole, C., Sobala, A., Lu, C., Thatcher, S. R., Bowman, A., Brown, J. W., Green, P. J., Barton, G. J., and Hutvagner, G. (2009) Filtering of deep sequencing data reveals the existence of abundant Dicer-dependent small RNAs derived from tRNAs. *RNA* **15**, 2147–2160.

Chapter 12

Computational Prediction of Plant miRNA Targets

Ying-Hsuan Sun, Shanfa Lu, Rui Shi, and Vincent L. Chiang

Abstract

MicroRNAs (miRNAs) are a specific class of 21-nt small RNAs. They regulate the expression of specific target genes by various types of post-transcriptional regulation mechanisms, such as transcript cleavage and translation suppression. The biological function of an miRNA is therefore intimately associated with the function of their target genes. Target gene identification becomes an essential step towards understanding miRNA functions. In this protocol, we describe a computational procedure for plant miRNA target prediction. It involves two key steps: (1) search of transcript sequence databases for target sequences that have a near-perfect sequence complementarity to the miRNA sequence using the "scan_for_matches" program and (2) evaluation of the miRNA:target sequence pair for pairing complementarity using specific rules, such as positional dependent penalty score and minimum free energy ratio filter, to identify the most likely candidate targets.

Key words: Computational prediction, microRNA, microRNA target, plant, small RNA.

1. Introduction

In plants, most of the interactions between an miRNA and its targets result in transcript-level reduction through mRNA cleavage mediated by the RNA-induced silencing complex (RISC) (1–3). These interactions involve guiding RISC to the mRNA target site through annealing to miRNAs. In animal systems, miRNAs and their targets are often bound less tightly, resulting in complicated miRNA:target duplex structures that are more difficult to predict through computational approaches (4, 5). The miRNA:target duplexes in plants are usually in near-perfect sequence complementation (6), allowing a more robust target prediction through computation. Current methods for the identification of miRNA targets are mainly based on computational

H. Kodama, A. Komamine (eds.), *RNAi and Plant Gene Function Analysis*, Methods in Molecular Biology 744,
DOI 10.1007/978-1-61779-123-9_12, © Springer Science+Business Media, LLC 2011

prediction. Validation of the targets is then achieved experimentally, such as by detection of the cleavage through the 5′-rapid amplification of cDNA ends (5′-RACE) method (3, 7), RNA degradome analysis (8, 9), or by genetic means.

The procedures for predicting an miRNA target can be divided into two steps: (1) finding miRNA's near-perfect complementary target sequences from the appropriate sequence sources and (2) evaluating the miRNA:target complementarity stringency. Many programs can perform homologous sequence search but not all can be applied in step 1. For example, BLAST (10) is the most commonly used program for homologous sequence search. It is based on a heuristic algorithm that is designed to find sequences that are most similar to the query. The result, however, does not encompass all possible homologous sequences. Besides, it is optimized for searching homologous sequences that are much longer than 21 bp, the length of an miRNA:target duplex. The program FASTA (11) is also commonly used for homologous sequence search. It can search short homologous sequences and has been applied for miRNA target prediction (12). However, like BLAST, it is also based on a heuristic algorithm and cannot identify all possible homologous sequences. Other programs, such as "scan_for_matches" (also known as PatScan, 13), used a brute force approach that tests all possible sequences to be searched and identifies ones that fit the searching criteria (such as number of insertion, deletion, and substitution allowed). It has been successfully applied to miRNA target prediction in plants (3, 7). Scan_for_matches does not suffer from the issue of sensitivity that BLAST or FASTA has, even though it is less efficient in terms of speed and requires more computational power. In order to achieve a robust target prediction, it is preferred to have all potential miRNA:target pairs identified for evaluation. In this protocol, we will use the program "scan_for_matches" to illustrate the miRNA target prediction process.

Following the identification of the miRNA:target duplexes, complementarity patterns of the duplex need to be evaluated for predicting the most likely targets. Many evaluation criteria were suggested. Rhoades et al. (6) predicted miRNA targets from the duplex formed with three or less mismatches. But this complementarity criterion was not sensitive enough to allow the prediction of many of the targets that were experimentally authenticated. Jones-Rhoades and Bartel (14) refined the criterion by including a penalty scoring system for mismatched patterns in the miRNA:target duplex within a 20-base sequence window. A 0 point was assigned to each complementary pair, 0.5 point to each G:U wobble, 1 point to each non-G:U wobble mismatch, and 2 points to each bulged nucleotide in either RNA strand. A score of 3.0 or less was demonstrated for predicting authentic targets (experimentally validated) with very high confidence. For example, a cutoff score of 3.0 can predict at least one authentic target for each conserved miRNA in *Arabidopsis* and rice. In *Populus trichocarpa*, some targets with a score of 3.0 have also been confirmed to be authentic based on 5′-RACE, while scores ≤ 2.5 effectively minimized the number of non-authentic targets (3). From a set of 94 authentic miRNA targets in *Arabidopsis*, Allen et al. (15) discovered that the *Arabidopsis* miRNA targets display a highly complementary region with the 5′-end of the miRNA. Based on these 94 authenticated miRNA targets in *Arabidopsis*, Allen et al. (15) further refined the penalty scoring system by doubling the penalty scores defined by Jones-Rhoades and Bartel (14), for all mismatch types appeared between the 2nd and the 13th nucleotides of the miRNA. A minimum free energy (MFE) ratio rule was also incorporated to further filter out false targets. Using a combination of penalty scores ≤ 4 and MFE ratio filters ≥ 0.73, Allen et al. (15) were able to correctly predict 87 of the 94 authentic miRNA targets.

Here we illustrate a stepwise computational protocol for plant miRNA target prediction, using principles described above. The computationally predicted targets need to be experimentally confirmed for their authenticity, such as by validation of target site cleavage through cloning and sequencing (3, 7–9).

2. Materials

2.1. A Sequence Database File

The sequence database is the source of putative miRNA targets to be searched. The file needs to be in the FASTA file format. A general FASTA file format description can be found from the National Center for Biotechnology Information (NCBI, http://www.ncbi.nlm.nih.gov/blast/fasta.shtml). An example file with predicted transcripts in the *P. trichocarpa* genome annotation v1.1 is shown in **Fig. 12.1**. This file is saved in this example with the filename "sequence_database" to demonstrate the procedures.

2.2. miRNA Sequences

The miRNA sequence in the reverse-complementary DNA format will be used for target search, while RNA sequence format will be used for displaying the miRNA:target alignment pattern.

2.3. A Computer System

A general PC or Apple computer system with at least 1.5-GHz CPU, 1 Mb of RAM, and 150 Gb of disk space should have enough capability to accomplish computational miRNA target prediction (*see* the example below). More memory, data storage, and faster CPU would be advantageous and may be necessary for a genome-wide scale prediction of miRNA targets.

```
>fgenesh1_pg.C_scaffold_219000006
ATGGTAGATTTGAAAGATACTCTTGCAAAGTTCACATCTGCTCTTAGTTTTCAGGAGAAATGTAAGTTTC
CATCTCAACCTCAACAAAATTCCAAAGGGCAATACAATTCAAGTGCGAGTAGCTCTAGAAGCCAACACAT
GGATCAAGTTAAATCAGTCACTACTCTTTGTCGTGGTAAGGTTATTGAAAAACCTATTCTTGAACCTTGT
AAGAAAGATGATGAGTTAATCTCTGAGGGTAAGGAAATGGTTGAACCTGAACATTGCAAAGAAAAGATTG
ATTTCCCACCAGTACTTCCATTTCCTAATGCCATGACCAAACAAAGAAAAGTCAATCACAATTCTGAAAT
CTTTGGAACTTTCAAACAGGAAAAATCAAAGGCAAATCAGCCCAACACCAAGCGTAATCATGCCAACGAC
CCATTGGAGGTGGCAATTGGGCCAATCACAAGAGCTAGGGCAAAGAAGCATTGA

>eugene3.02190008
ATGTGGGGAATTGTTTTTTACCACCAACCAATTCTTGCAGCTACTATCATGTCAGATCAAAAGAACAAAA
CACCCCCTCCCAATGTAGAACTTAAATTCCAAATGCAAGCCATGACGAAGATGATGGAAAGAATGAATTC
CGTGATGGGGAATGTGTGTGACAGACTTGAGAAAGTGGGAAAACAAGGTAATGTCAGAACATGTACCCAA
GACGTGAGAAAGGTTGGGGCTGAACCAAAATCAAACAATGGCAGAGGGGCTGAAAGGCCAAGGTGGGCTG
ATTATGCGGATTTTGAGGTGGACGTTGATGATATTGTTGATGGTGGTTTTAAGGATGAGACCATAGGCCA
TCAAAAAGGTTTTCAACACCATAGAAACCGAAGGGATTTCATGTATTTTGACGGGGTGTTATGGCAAAAG
AAAATGAGGATTCAAAAGGAGAGGTGTCAAAGGGAGAGAAATAAAGAGATTGGTGTTCTAAAAAATGAAT
CCAAGAGTCTATACCGTATTCTAGGGGAGATGAAGCAAGAACTTGATGTGTTAATGGCAATAGTCAATGC
CCAACAAGCTTCAGAAAGAGAGGAGAAAATACGCTGCAATGATGATATACGAAATAGAATGGATGCTACT
TTCATCAAAAGTTGGTGGCGACGATTGTTGGCAAGAAAAGAACTCCAAAGGCTTCAAAAAGAGGCTAAGg
aatttggtccttaa

>eugene3.02190007
ATGGAAGGCATGGCTAGTTCAGCAGCACAAACAACCCCTACCACGCAGATTCCACCAGCAGCCCCTACGA
CCAGCATGACCATGGATAATGTGGTACCACTGGTGCGACTAGTCAAGAGTATGAGGGAAATGGGCTGTGA
ACCGTATATGGGGGAACAAGATGCAGAGATAGCTGGAAGATTGATCAGGAAGGTAGAAAAGACAATGATT
.
.
.
```

Fig. 12.1. A sequence database file in FASTA file format.

2.4. Softwares

1. Program "scan_for_matches". This program is used for searching miRNA sequence patterns in the sequence database. The program source code can be downloaded from http://iubio.bio.indiana.edu/soft/molbio/pattern/ or http://biopieces.googlecode.com/files/scan_for_matches.tar.gz, and needs to be compiled as instructed by the program manual.
2. Program "RNAhybrid" (16). This program is used for calculating minimum free energy ratio of the double-stranded target region and miRNA:target duplex as described by Allen et al. (15). "RNAhybrid" can be downloaded from http://bibiserv.techfak.uni-bielefeld.de/download/tools/rnahybrid.html. Program source code and executables compiled for both Windows and Mac operating systems are available from the download site.
3. A simple text-editing program. Programs such as "notepad" in the Windows operating system or "TextEdit" in the Mac

operating system are needed for creating and editing files used in the computational target prediction procedure.

4. PERL scripting language. It is used to run scripts for evaluating the miRNA:target complementary pattern. PERL should be pre-installed in the operating systems, such as Linux or Mac OS. A version of PERL for the Windows operating system can be downloaded from http://www.activestate.com/activeperl/.

3. Methods

3.1. Searching for Target Sequences with Near-Perfect Complementarity to the miRNA Sequence

The search for target sequences containing near-perfect complementarity to the miRNA by the program "scan_for_matches" is invoked with the following simple syntax:

scan_for_matches pattern_file < sequence_database > output_file

Scan_for_matches is a program capable of searching fuzzy and complicated sequence patterns by setting up options when executing the program and defining rules in pattern_file (details can be found in the "README" file of the package). For searching putative miRNA targets, rules for mismatches, deletion, and insertion are needed. These rules are set in the pattern_files as described below:

1. Use the text-editing program to create pattern_files based on miRNA sequences. A total of three pattern_files are required for the search. Each file contains a single line of text that starts with the reverse-complemented miRNA sequence in the DNA sequence format, followed by a bracket that contains three numbers separated by a comma. Each of these numbers represents the maximum number for mismatches, insertions, and deletions allowed for the search, respectively. The three pattern_files are used to complete the search of sequence patterns that are complementary to miRNA sequences with up to seven mismatches, up to three mismatches plus one insertion, and up to three mismatches plus one deletion, respectively. The search for one insertion and one deletion patterns allows accommodation of single-nucleotide "bulge" patterns in the miRNA:target pair.

   As an example, the procedure for searching the targets of *P. trichocarpa* miRNA156a (ptc-miR156a) is illustrated in **Fig. 12.2**. Pattern_file "ptc-miR156a.700" allows the search of sequence patterns with up to seven mismatches, with no insertion or deletion (**Fig. 12.2a**). Pattern_file "ptc-miR156a.310" allows up to three mismatches plus one insertion (**Fig. 12.2b**). Pattern_file "ptc-miR156a.301" allows up to three mismatches plus one deletion (**Fig. 12.2c**).

```
A GTGCTCACTCTCTTCTGTCA[7,0,0]

B GTGCTCACTCTCTTCTGTCA[3,1,0]

C GTGCTCACTCTCTTCTGTCA[3,0,1]
```

Fig. 12.2. Contents of pattern_files for "scan_for_matches" search. **a** Contents of file "ptc-miR156a.700". No more than seven mismatches are allowed, while the pattern contains no insertion or deletion. **b** Contents of file "ptc-miR156a.310". No more than three mismatches plus one insertion are allowed. **c** Contents of file "ptc-miR156a.301". No more than three mismatches plus one deletion are allowed.

A
```
[prompt]% /path_to_program/scan_for_matches ptc-miR156a.700 < sequence_database > ptc-miR156a_700.match
[prompt]%
```

B
```
[prompt]% /path_to_program/scan_for_matches ptc-miR156a.310 < sequence_database > ptc-miR156a_310.match
[prompt]%
```

C
```
[prompt]% /path_to_program/scan_for_matches ptc-miR156a.301 < sequence_database > ptc-miR156a_301.match
[prompt]%
```

Fig. 12.3. The commands for invoking ptc-miR156a sequence pattern search by program "scan_for_matches". **a** Searching for sequence patterns containing up to seven mismatches while no insertion or deletion allowed and saving results to file "ptc-miR156a_700.match". **b** Searching for sequence patterns containing up to three mismatches plus one insertion and saving results to file "ptc-miR156a_310.match". **c** Searching for sequence pattern containing up to three mismatches plus one deletion and saving results to file "ptc-miR156a_301.match".

2. Execute commands to invoke the sequence pattern search. At the prompt of the command line interface, type in the scan_for_matches commands, as illustrated in **Fig. 12.3a–c**. Replace "path_to_program" with the actual directory where the program "scan_for_matches" is located.

   In the example shown in **Fig. 12.3a–c**, each pattern is separately executed. The search results are saved in the files "ptc-miR156a_700.match" (**Fig. 12.3a**), "ptc-miR156a_310.match" (**Fig. 12.3b**), and "ptc-miR156a_301.match" (**Fig. 12.3c**), respectively. The result output is in the FASTA format as shown in **Fig. 12.4**, in which search output file "ptc-miR156a_700.match" is set as an example. Each hit record contains two lines. The header line starts with ">" and is followed by the sequence ID and a bracket, in which the start and end locations of the match are indicated. The second line shows the matched sequence pattern.

3.2. Evaluate the Penalty Score

Calculate the penalty score of each matched pattern. A comprehensive penalty score matrix was given by Jones-Rhoades and

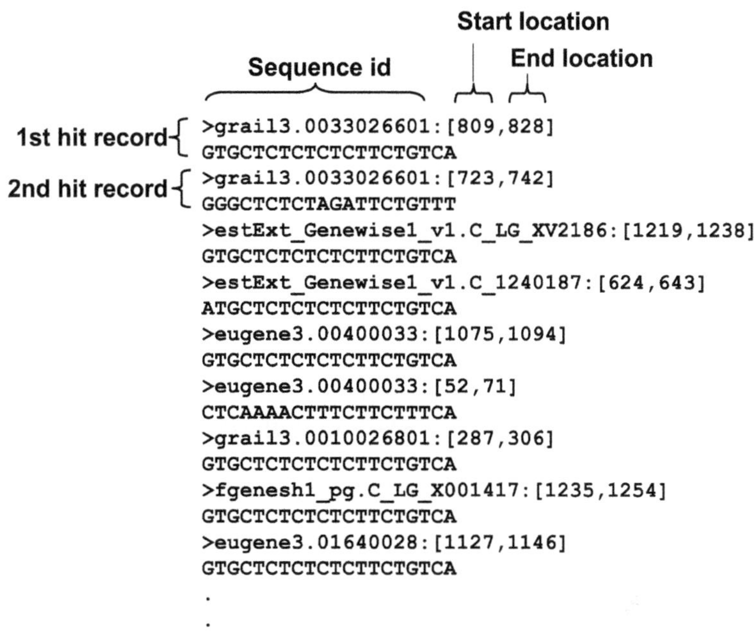

Fig. 12.4. Contents of the "scan_for_matches" result file, "ptc-miR156a_700.match". Each matched hit record has two lines, with the header line started with ">" followed by the sequence ID and the coordinate of the hit. The second line is the matched target sequence.

Bartel (14). The scoring system calculates the mismatched patterns in the miRNA:target duplexes within a 20-base sequence window, with 0 point being assigned to each complementary pair, 0.5 point to each G:U pair, 1 point to each non-G:U mismatch, and 2 points to each bulged nucleotide (insertion or deletion) in either RNA strand. Thus, the penalty score of an miRNA:target pair equals (1 × the number of non-G:U mismatch)+(2 × the number of bulged nucleotide)+(0.5 × the number of G:U pair). For an miRNA longer than 20 nt, the lowest score of the 20-base sequence window is assigned as the penalty score of the pair. Allen et al. (15) showed that, counting from the 3′-end of the target, nucleotides at the positions 2–13 of an miRNA:target pair contained relatively few mismatches as compared to those at the positions 1 and 14–21. A modified scoring scheme was then devised by doubling the penalty score for all mismatches located from positions 2 to 13. Allen et al. (15) call this score system "position-dependent scoring system" and it is therefore termed "position-dependent penalty score" in this protocol.

To illustrate the calculation of the penalty score, the first hit record in the search result file "ptc-miR156a.700" is used as an example. The first matched target is located from 809 to 828 nt of the gene grail3.0033026601 (**Fig. 12.5a**). The complementary sequence pattern of the predicted target

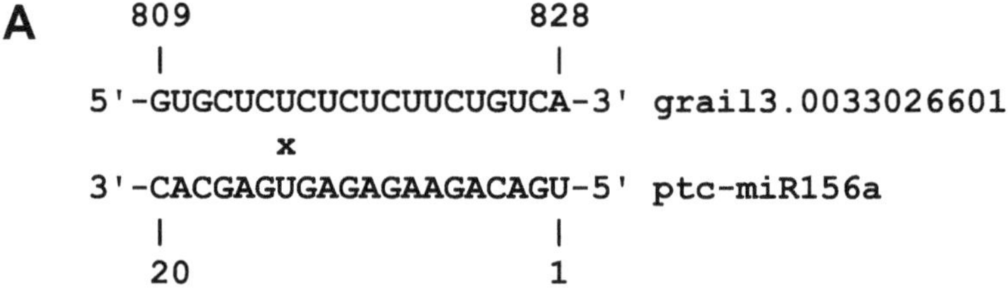

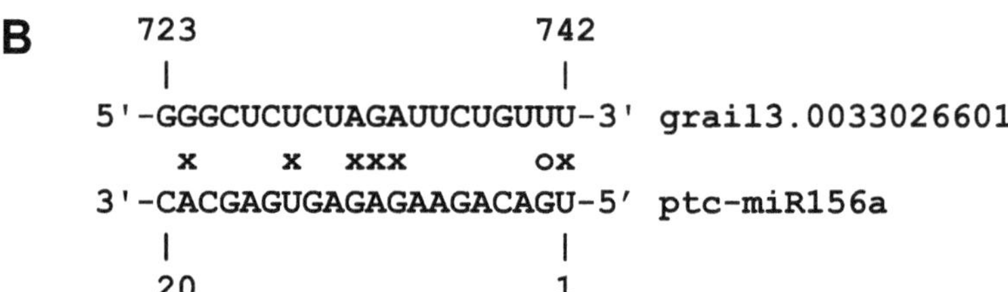

Fig. 12.5. Complementary pattern of the ptc-miR156a and the target region of the first two hit records from the "scan_for_matches" result file "ptc-miR156a_700.match". **a** Sequence alignment of the ptc-miR156a and the target region of the first hit record. **b** Sequence alignment of the ptc-miR156a and the target region of the second hit record.

region and ptc-miR156a shows one mismatch, which is located at the 19 nt of ptc-miR156a. Therefore, the penalty score is $(1\times1)+(2\times0)+(0.5\times0)=1$. Because no mismatch is found between positions 2 and 13 nt of ptc-miR156a, the position-dependent penalty score is also equal to 1.

A second hit is also found in grail3.0033026601, 723–742 nt (**Fig. 12.5b**). The complementary pattern of the duplex contains six mismatches and one G:U pair (**Fig. 12.5b**), giving a penalty score of $(1\times6)+(2\times0)+(0.5\times1) = 6.5$. Because this score is greater than the cutoff value of 3 defined by Jones-Rhoades and Bartel (14), this second hit can be excluded as one that contains a target cleavage site. In addition, in this second duplex, three of the six mismatches and the G:U pair are located within the positions 2–13 nt of ptc-miR156a and three mismatches located on the first or 14–21 nt of ptc-miR156a, affording a position-dependent penalty score of $2\times\ ((1\times3)+(2\times0)+(0.5\times1))+((1\times3)+(2\times0)+(0.5\times0)) = 10$. Similarly, this value, which is greater than the cutoff score of 4 defined by Allen et al. (15), disqualifies this second hit as a target sit. Therefore, based on both scoring systems, ptc-miR156a's predicted target is the gene grail3.0033026601 with a likely cleavage site between 809 and 828 nt of the gene.

3.3. Evaluate the Minimum Free Energy (MFE) Ratio

Calculate MFE ratio for the putative target that passes the penalty score filter. Allen et al. (15) showed that the MFE ratio ($\Delta G_{target}/\Delta G_{MFE}$) could be used to further reduce the false rate of target prediction that is based on the penalty score. To calculate the MFE ratio, the matched target site sequence from search results was used. ΔG_{MFE} is calculated for the hypothetical double-stranded RNA sequences containing the miRNA target site region, while ΔG_{target} is calculated for the actual miRNA:target duplex.

Allen et al. (15) used the program "RNAfold" in the Vienna package for MFE calculation. First, a pseudo-RNA molecule is created by inserting a 4-base gap linker sequence between an miRNA and its corresponding mRNA target site and between an mRNA target site and its reverse complementary form. However, it is possible that this inserted linker may participate in stem formation rather than the loop as it is originally intended and render the MFE value inaccurate. Other programs, such as "RNAhybrid" and "hybrid-min" in the UNAfold package ((17), http://dinamelt.bioinfo.rpi.edu/download.php) can be used to calculate MFE for two independent RNA sequences without creating an artificial linker. In the following example, we use the program "RNAhybrid" for MFE ratio calculation for ptc-miR156a and its first hit record from ptc-miR156a_700.match. The calculation procedures are listed below:

1. Use the text-editing program to create a ptc-miR156a.seq file that contains the ptc-miR156a RNA sequence in FASTA format as listed below:

   >ptc-miR156a

   UGACAGAAGAGAGUGAGCAC

2. Extract the first matched target sequence from ptc-miR156a_700.match and save in RNA FASTA format as T.seq:

   >T.seq

   GUGCUCUCUCUCUUCUGUCA

3. Prepare a sequence that is reversely complementary to T.seq in the FASTA format and save it as Trc.seq:

   >Trc.seq

   UGACAGAAGAGAGAGAGCAC

4. Calculate ΔG_{MFE} for T.seq and Trc.seq by executing the command listed below in the command line interface. *Note*: replace the "Path_to_program" with the actually directory where the program is located:

   [prompt]% /Path_to_program/RNAhybrid –t T.seq –q Trc.seq –d theta

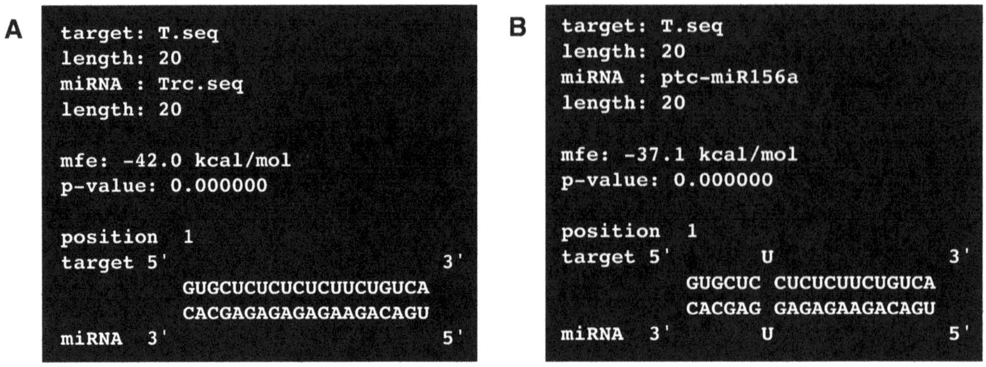

A
```
target: T.seq
length: 20
miRNA : Trc.seq
length: 20

mfe: -42.0 kcal/mol
p-value: 0.000000

position  1
target 5'                    3'
       GUGCUCUCUCUCUUCUGUCA
       CACGAGAGAGAGAAGACAGU
miRNA  3'                    5'
```

B
```
target: T.seq
length: 20
miRNA : ptc-miR156a
length: 20

mfe: -37.1 kcal/mol
p-value: 0.000000

position  1
target 5'        U           3'
       GUGCUC CUCUCUUCUGUCA
       CACGAG GAGAGAAGACAGU
miRNA  3'        U           5'
```

Fig. 12.6. The outputs of minimum free energy $\triangle G_{MFE}$ and $\triangle G_{target}$ calculated by the program "RNAhybrid". **a** $\triangle G_{MFE}$ calculation result. **b** $\triangle G_{target}$ calculation result.

The $\triangle G_{MFE}$ is –42 as shown in the result output (**Fig. 12.6a**).

5. Calculate $\triangle G_{target}$ for T.seq and ptc-miR156a.seq by executing the command listed below in the command line interface. *Note*: replace the "Path_to_program" with the actually directory where the program is located:

   [prompt]% /Path_to_program/RNAhybrid –t T.seq –q ptc-miR156a.seq –d theta

   The $\triangle G_{target}$ is −37.1 as shown in the result output (**Fig. 12.6b**).

6. Thus, MFE ratio = $\triangle G_{target} / \triangle G_{MFE}$ = (−37.1)/(–42) = 0.88, which is greater than 0.73 as suggested by Allen et al. (15). The result of MFE ratio further strengthens the prediction that grail3.0033026601 is a target of ptc-miR156a with a cleavage site between 809 and 828 nt of the target gene.

3.4. Conclusion

The procedure described above shows the process of analyzing one putative target gene for an miRNA. In real life, one may have a project that requires predicting targets for hundreds of miRNAs. To do that, a computer program to automate the whole process is needed. We have developed a program named "target_prediction" using PERL language and the above-described "scan_for_matches" and "RNAhybrid" programs to automate and streamline the target prediction process for a large number of miRNAs. Our program takes a list of miRNA sequences and searches for near-perfect complements in the database, evaluates the miRNA:target complement patterns, and outputs the prediction results in an easy-to-read html web table format (**Fig. 12.7**). The program is available for download from http://www.ncsu.edu/forestbiotech/miRNA/target_prediction.

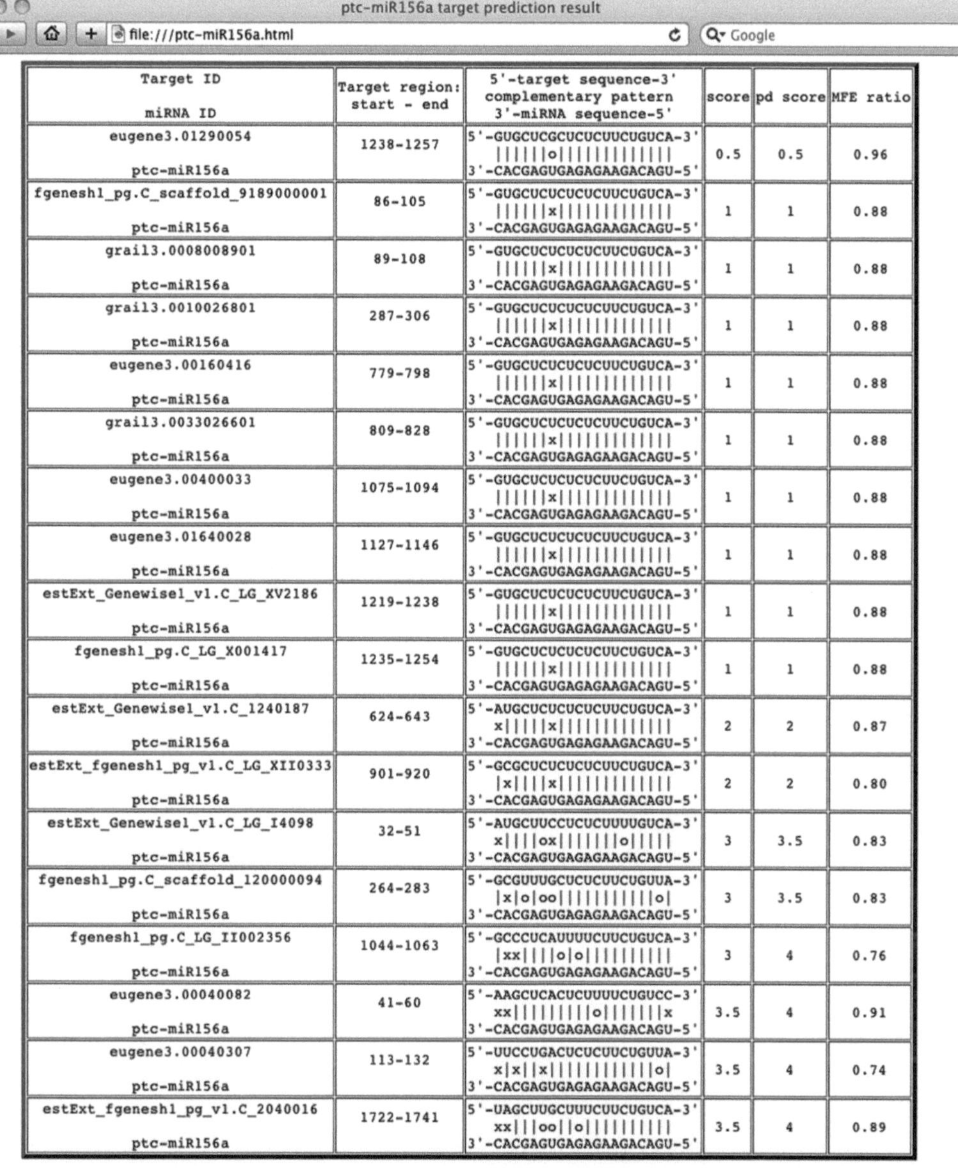

<table>
<tr><th>Target ID
miRNA ID</th><th>Target region: start - end</th><th>5'-target sequence-3'
complementary pattern
3'-miRNA sequence-5'</th><th>score</th><th>pd score</th><th>MFE ratio</th></tr>
<tr><td>eugene3.01290054
ptc-miR156a</td><td>1238-1257</td><td>5'-GUGCUCGCUCUCUUCUGUCA-3'
||||||o|||||||||||||
3'-CACGAGUGAGAGAAGACAGU-5'</td><td>0.5</td><td>0.5</td><td>0.96</td></tr>
<tr><td>fgenesh1_pg.C_scaffold_9189000001
ptc-miR156a</td><td>86-105</td><td>5'-GUGCUCUCUCUCUUCUGUCA-3'
||||||x|||||||||||||
3'-CACGAGUGAGAGAAGACAGU-5'</td><td>1</td><td>1</td><td>0.88</td></tr>
<tr><td>grail3.0008008901
ptc-miR156a</td><td>89-108</td><td>5'-GUGCUCUCUCUCUUCUGUCA-3'
||||||x|||||||||||||
3'-CACGAGUGAGAGAAGACAGU-5'</td><td>1</td><td>1</td><td>0.88</td></tr>
<tr><td>grail3.0010026801
ptc-miR156a</td><td>287-306</td><td>5'-GUGCUCUCUCUCUUCUGUCA-3'
||||||x|||||||||||||
3'-CACGAGUGAGAGAAGACAGU-5'</td><td>1</td><td>1</td><td>0.88</td></tr>
<tr><td>eugene3.00160416
ptc-miR156a</td><td>779-798</td><td>5'-GUGCUCUCUCUCUUCUGUCA-3'
||||||x|||||||||||||
3'-CACGAGUGAGAGAAGACAGU-5'</td><td>1</td><td>1</td><td>0.88</td></tr>
<tr><td>grail3.0033026601
ptc-miR156a</td><td>809-828</td><td>5'-GUGCUCUCUCUCUUCUGUCA-3'
||||||x|||||||||||||
3'-CACGAGUGAGAGAAGACAGU-5'</td><td>1</td><td>1</td><td>0.88</td></tr>
<tr><td>eugene3.00400033
ptc-miR156a</td><td>1075-1094</td><td>5'-GUGCUCUCUCUCUUCUGUCA-3'
||||||x|||||||||||||
3'-CACGAGUGAGAGAAGACAGU-5'</td><td>1</td><td>1</td><td>0.88</td></tr>
<tr><td>eugene3.01640028
ptc-miR156a</td><td>1127-1146</td><td>5'-GUGCUCUCUCUCUUCUGUCA-3'
||||||x|||||||||||||
3'-CACGAGUGAGAGAAGACAGU-5'</td><td>1</td><td>1</td><td>0.88</td></tr>
<tr><td>estExt_Genewise1_v1.C_LG_XV2186
ptc-miR156a</td><td>1219-1238</td><td>5'-GUGCUCUCUCUCUUCUGUCA-3'
||||||x|||||||||||||
3'-CACGAGUGAGAGAAGACAGU-5'</td><td>1</td><td>1</td><td>0.88</td></tr>
<tr><td>fgenesh1_pg.C_LG_X001417
ptc-miR156a</td><td>1235-1254</td><td>5'-GUGCUCUCUCUCUUCUGUCA-3'
||||||x|||||||||||||
3'-CACGAGUGAGAGAAGACAGU-5'</td><td>1</td><td>1</td><td>0.88</td></tr>
<tr><td>estExt_Genewise1_v1.C_1240187
ptc-miR156a</td><td>624-643</td><td>5'-AUGCUCUCUCUCUUCUGUCA-3'
x|||||x|||||||||||||
3'-CACGAGUGAGAGAAGACAGU-5'</td><td>2</td><td>2</td><td>0.87</td></tr>
<tr><td>estExt_fgenesh1_pg_v1.C_LG_XII0333
ptc-miR156a</td><td>901-920</td><td>5'-GCGCUCUCUCUCUUCUGUCA-3'
|x||||x|||||||||||||
3'-CACGAGUGAGAGAAGACAGU-5'</td><td>2</td><td>2</td><td>0.80</td></tr>
<tr><td>estExt_Genewise1_v1.C_LG_I4098
ptc-miR156a</td><td>32-51</td><td>5'-AUGCUUCCUCUCUUUUGUCA-3'
x||||ox|||||||o|||||
3'-CACGAGUGAGAGAAGACAGU-5'</td><td>3</td><td>3.5</td><td>0.83</td></tr>
<tr><td>fgenesh1_pg.C_scaffold_120000094
ptc-miR156a</td><td>264-283</td><td>5'-GCGUUUGCUCUCUUCUGUUA-3'
|x|o|oo|||||||||||o|
3'-CACGAGUGAGAGAAGACAGU-5'</td><td>3</td><td>3.5</td><td>0.83</td></tr>
<tr><td>fgenesh1_pg.C_LG_II002356
ptc-miR156a</td><td>1044-1063</td><td>5'-GCCCUCAUUUUCUUCUGUCA-3'
|xx||||o|o||||||||||
3'-CACGAGUGAGAGAAGACAGU-5'</td><td>3</td><td>4</td><td>0.76</td></tr>
<tr><td>eugene3.00040082
ptc-miR156a</td><td>41-60</td><td>5'-AAGCUCACUCUUUUCUGUCC-3'
xx|||||||||o|||||||x
3'-CACGAGUGAGAGAAGACAGU-5'</td><td>3.5</td><td>4</td><td>0.91</td></tr>
<tr><td>eugene3.00040307
ptc-miR156a</td><td>113-132</td><td>5'-UUCCUGACUCUCUUCUGUUA-3'
x|x||x||||||||||||o|
3'-CACGAGUGAGAGAAGACAGU-5'</td><td>3.5</td><td>4</td><td>0.74</td></tr>
<tr><td>estExt_fgenesh1_pg_v1.C_2040016
ptc-miR156a</td><td>1722-1741</td><td>5'-UAGCUUGCUUUCUUCUGUCA-3'
xx|||oo||o||||||||||
3'-CACGAGUGAGAGAAGACAGU-5'</td><td>3.5</td><td>4</td><td>0.89</td></tr>
</table>

Fig. 12.7. The web page table output of target prediction results using the program "target_prediction". The table has six columns, which contain information of the ID of target and miRNA, target region, complementary pattern, penalty score (score), position-dependent penalty score (pd score), and MFE ratio, respectively. In the column with complementary pattern, "|" represents a matched base pair, "x" represents a non-G:U mismatch, and "o" represents a G:U wobble mismatch.

References

1. Kidner, C. A. and Martienssen, R. A. (2005) The developmental role of microRNA in plants. *Curr. Opin. Plant Biol.* **8**, 38–44.
2. Llave, C., Xie, Z., Kasschau, K. D., and Carrington, J. C. (2002) Cleavage of Scarecrow-like mRNA targets directed by a class of *Arabidopsis* miRNA. *Science* **297**, 2053–2056.

3. Lu, S., Sun, Y. H., Shi, R., Clark, C., Li, L., and Chiang, V. L. (2005) Novel and mechanical stress-responsive microRNAs in *Populus trichocarpa* that are absent from *Arabidopsis*. *Plant Cell* **17**, 2186–2203.
4. Grosshans, H. and Slack, F. J. (2002) MicroRNAs: Small is plentiful. *J. Cell. Biol.* **156**, 17–21.
5. Lewis, B. P., Shih, I., Jones-Rhoades, M. W., Bartel, D. P., and Burge, C. B. (2003) Prediction of mammalian microRNA targets. *Cell* **115**, 787–798.
6. Rhoades, M. W., Reinhart, B. J., Lim, L. P., Burge, C. B., Bartel, B., and Bartel, D. P. (2002) Prediction of plant microRNA targets. *Cell* **110**, 513–520.
7. Lu, S., Sun, Y. H., and Chiang, V. L. (2008) Stress-responsive microRNAs in *Populus*. *Plant J.* **55**, 131–151.
8. Addo-Quaye, C., Eshoo, T. W., Bartel, D. P., and Axtell, M. J. (2008) Endogenous siRNA and miRNA targets identified by sequencing of the *Arabidopsis* degradome. *Curr. Biol.* **18**, 758–762.
9. German, M. A., Pillay, M., Jeong, D. H., Hetawal, A., Luo, S., Janardhanan, P., Kannan, V., Rymarquis, L. A., Nobuta, K., German, R., De Paoli, E., Lu, C., Schroth, G., Meyers, B. C., and Green, P. J. (2008) Global identification of microRNA–target RNA pairs by parallel analysis of RNA ends. *Nat. Biotechnol.* **26**, 941–946.
10. Altschul, S. F., Gish, W., Miller, W., Myers, E. W., and Lipman, D. J. (1990) Basic local alignment search tool. *J. Mol. Biol.* **215**, 403–410.
11. Pearson, W. R., and Lipman, D. J. (1988) Improved tools for biological sequence comparison. *Proc. Natl. Acad. Sci. USA* **85**, 2444–2448.
12. Fahlgren, N. and Carrington, J. C. (2009) MiRNA target prediction in plants. In: *Plant microRNAs Methods and Protocols* (eds. Meyers, B. C. and Pamela, J. G.), Springer, Berlin, pp. 19–30.
13. Dsouza, M., Larsen, N., and Overbeek, R. (1997) Searching for patterns in genomic data. *Trends Genet.* **13**, 497–498.
14. Jones-Rhoades, M. W. and Bartel, D. P. (2004) Computational identification of plant microRNAs and their targets, including a stress-induced miRNA. *Mol. Cell* **14**, 787–799.
15. Allen, E., Xie, Z., Gustafson, A. M., and Carrington, J. C. (2005) MicroRNA-directed phasing during *trans*-acting siRNA biogenesis in plants. *Cell* **121**, 207–221.
16. Rehmsmeier, M., Steffen, P., Höchsmann, M., and Giegerich, R. (2004) Fast and effective prediction of microRNA/target duplexes. *RNA* **10**, 1507–1517.
17. Markham, N. R. and Zuker, M. (2008) UNAFold: Software for nucleic acid folding and hybridization. In: *Bioinformatics, Volume II. Structure, Functions and Applications*, no. 453, *Methods in Molecular Biology* (eds. Keith, J. M.), Humana Press, Totowa, NJ, pp. 3–31.

Chapter 13

Bisulfite Sequencing for Cytosine-Methylation Analysis in Plants

Nazmul Haque and Masamichi Nishiguchi

Abstract

RNA silencing is a sequence-specific RNA degradation mechanism conserved in eukaryotes including fungi, plants, and animals. One of the three RNA silencing pathways is DNA methylation which is the result of interaction between DNA and siRNA, a hallmark of RNA silencing. Bisulfite sequencing can be very powerful for DNA methylation analysis in this context. This method includes DNA extraction, digestion of DNA with restriction enzyme, treatment of DNA with bisulfite, PCR amplification of DNA, cloning of amplified DNA fragments, sequencing of DNA fragments, and analysis of DNA sequences. Based on this method, increased levels of cytosine methylation were obtained in both symmetrical (CpG, CpNpG) and non-symmetrical (CpHpH) contexts in silenced lines of transgenic plants (CG>CNG>CHH), while the methylation levels were low in nonsilenced, over-expressing lines. Through grafting, RNA silencing was induced in the non-silenced scions from silenced rootstocks; however, the methylation level of DNA in the scions did not increase.

Key words: Bisulfite, cytosine, DNA methylation, RNA silencing, transgene.

1. Introduction

RNA silencing, a general term used to describe post-transcriptional gene silencing in plants, also called quelling in fungi and RNA interference in animals, is a highly conserved process of sequence-specific RNA degradation observed in almost all eukaryotes examined to date. Molecular analyses of transgenic plants reveal a close association between RNA silencing and DNA methylation of the transgene (1), suggesting that DNA methylation may have a role in triggering or maintaining the RNA silencing pathway or both.

H. Kodama, A. Komamine (eds.), *RNAi and Plant Gene Function Analysis*, Methods in Molecular Biology 744,
DOI 10.1007/978-1-61779-123-9_13,

Methylation of cytosine in DNA is important for gene activation/inactivation, including RNA silencing, in various organisms. For methylation analysis, several methods are available. One choice is bisulfite sequencing, a characteristic of which is its comprehensive analysis of cytosine methylation in the DNA region of interest, while other methods, including methylation-specific restriction analysis and methylated DNA immunoprecipitation, provide only limited information even in a very short region of DNA. Bisulfite sequencing is based on the different effects of sodium bisulfite on methylated/non-methylated cytosine (2–4). Sodium bisulfite induces deamination of cytosine base followed by conversion of deaminated cytosine into uracil. However, the amount of conversion into uracil is extremely low for methylated cytosine. That is, methylated cytosine remains unchanged, while unmethylated cytosine is changed into uracil. Thus methylation analysis based on this method consists mainly of three parts: bisulfite treatment of DNA, PCR amplification using appropriate primers common to methylated/non-methylated DNA followed by cloning of PCR fragments, and sequencing of the clones.

Here, we describe DNA methylation analysis using the bisufite sequencing method in RNA-silenced/non-silenced lines of transgenic *Nicotiana benthamiana* plants carrying coat protein (*CP*) gene of sweet potato feathery mottle virus (SPFMV) (5, 6) This chapter describes the method we used from DNA preparation through DNA sequence analysis.

2. Materials

2.1. Plant Material and Genomic DNA Extraction

1. MS medium with 3% sucrose (MP Biomedicals, Solon, CA, USA).
2. DNeasy Plant Mini Kit including buffers AP1, AP2, AP3/E, AE, and AW (Qiagen, Hilden, Germany). QIAshredder mini spin column, DNeasy Mini spin column, and RNase are provided in this kit.

2.2. Restriction Digestion of Genomic DNA

1. Appropriate restriction endonuclease having no restriction sites within the gene of interest or the region of gene to be analyzed.
2. Phenol–chloroform–isoamyl alcohol (25:24:1).
3. Ice-cold absolute and 70% ethanol.
4. 3 M sodium acetate, pH 5.0.

2.3. Bisulfite Treatment of Genomic DNA

1. EZ DNA Methylation-Gold Kit including CT conversion reagent, M-dilution, M-dissolving, M-binding, M-wash,

M-desulfonation, and M-elution buffers. Zymo-spin IC column is also supplied in this kit (Zymo Research, Orange, CA, USA).

2.4. PCR Amplification of Bisulfite-Treated DNA and Gel Cleanup

1. Bisulfite-treated DNA.
2. Specific primers for the region of interest.
3. Mg^{2+}-free, 10× PCR buffer (Sigma Aldrich, St. Louis, MO, USA).
4. dNTPs (2 mM each, NEB, Ipswich, MA, USA).
5. 25 mM $MgCl_2$.
6. JumpStart Taq DNA polymerase (Sigma Aldrich, St. Louis, MO, USA).
7. 50× TAE stock: 2 M Tris, 1 M acetate, 100 mM Na_2EDTA, pH 8.0.
8. Low-melting agarose (NuSieve GTG; Cambrex, USA): Made 2% gel with 1× TAE buffer.
9. QIAquick Gel Extraction Kit (Qiagen, Valencia, CA, USA).

2.5. Cloning and Screening of Amplified DNA Sequences

1. pGEM-T Easy Vector System (Promega, Madison, WI, USA).
2. High-efficiency competent cells, JM109 (Promega, Madison, WI, USA).
3. LB plates with 100 μg/mL ampicillin.
4. SOC medium: 2% tryptone, 0.5% yeast extract, 10 mM NaCl, 20 mM glucose, 10 mM $MgCl_2$, 10 mM $MgSO_4$, 2.5 mM KCl, pH 7.5.
5. 100 mM Isopropyl-β-D-thiogalactopyranoside (IPTG).
6. 50 mg/mL 5-Bromo-4-chloro-3-indolyl-β-D-galactoside (X-Gal).
7. QIAprep Spin Miniprep Kit (Qiagen, Valencia, CA, USA).
8. Restriction endonuclease *Eco*RI.

2.6. Sequencing of Clones

1. Bigdye Terminal Cycle Sequencing Kit (Applied Biosystems, Foster City, CA, USA).

3. Methods

After the first demonstration by Frommer et al. (7), bisulfite genome sequencing has become one of the most widely used techniques to detect 5-methylcytosine residues (5 mC) in DNA at a single nucleotide resolution. Though robust, bisulfite genome sequencing is associated with some technical difficulties and potential artifacts. However, a standard protocol for bisulfite reaction has been established as well as specialized modifications for precise starting material, all based on the following fundamental procedure: sheering or fragmentation, denaturation, bisulfite conversion, desulfonation of genomic DNA, and amplification of DNA of interest with specific primers followed by cloning and sequencing.

The efficiency of this technique depends on the complete conversion of unmethylated cytosines to uracils as well as complete non-conversion of 5 mC. The accuracy of the data obtained might be significantly compromised if certain steps in this method are not performed with care. In particular, high-quality clean genomic DNA as starting material is a prerequisite in obtaining reliable and reproducible data. The presence of minute amounts of protein in DNA samples is detrimental to complete bisulfite conversion, which eventually leads to a higher level of erroneous DNA methylation. Another issue in the standard methodology is DNA degradation, an undesired side effect of bisulfite treatment that considerably limits downstream PCR efficiency. In fact, 84–96% of DNA was found to be degraded in a conventional 4–18 h bisulfite conversion reaction (8). In the same way, several other factors, including the amount of starting material, bisulfite treatment condition, and PCR condition, can lead to incorrect or non-reproducible results. Considering all these steps, we optimized the whole method starting with the extraction of genomic DNA from plant samples to the final product, clones ready for sequencing reactions. In this method, we used several commercially available kits that efficiently obtain accurate and reproducible data.

Here, we describe DNA methylation analysis in transgenic *N. benthamiana* plants carrying coat protein (*CP*) gene of sweet potato feathery mottle virus (SPFMV) (5, 6).

3.1. Preparation of Genomic DNA with DNeasy Plant Mini Kit

1. Plant seedlings, initially grown on MS media with 3% sucrose under sterile conditions, are transferred to soil and the genomic DNA is prepared from approximately 2- to 4-week-old plants using the DNeasy Plant Mini Kit (Qiagen) with some modifications.
2. Fresh plant leaves (120–150 mg) are collected and rinsed with bidistilled water.
3. Leaves are grounded to a fine powder in the presence of liquid nitrogen using a precooled mortar and pestle and subsequently transferred to a 2-mL microtube.
4. A 400 μL of buffer AP1 and 4 μL of 100 mg/mL RNase A mixture are added to the tube (*see* **Note 1**), then vortexed

vigorously for 1 min or until any tissue clump is resuspended (*see* **Note 2**).

5. Cells are lysed by incubating for 10 min at 65°C. The lysate is mixed three times during incubation by inverting. Then 130 μL of the AP2 buffer is added to the lysate, mixed, and incubated on ice for 5 min.
6. Lysate is centrifuged for 5 min at 20,000×*g* at room temperature (RT) (*see* **Note 3**).
7. The supernatant is transferred into the QIAshredder mini spin column placed in a 2-mL collection tube and centrifuged for 2 min at 20,000×*g* (*see* **Note 4**).
8. Flow-through is transferred from the collection tube into a new tube (*see* **Note 5**).
9. The volume is determined using a micropipette.
10. AP3/E buffer (1.5× volume of the flow-through in step 9) is directly added to the cleared lysate and mixed immediately by pipetting (*see* **Note 6**).
11. An aliquot of 650 μL of the mixture is transferred into the DNeasy Mini spin column placed in a 2-mL tube, centrifuged for 1 min at 6,000×*g*, and flow-through is discarded. This step is repeated with the remaining sample.
12. The DNeasy Mini spin column is placed into a new 2-mL tube and 500 μL buffer AW is added, centrifuged for 1 min at 6,000×*g*, and the supernatant discarded.
13. Another 500 μL of buffer AW is added to the DNeasy Mini spin column and centrifuged for 2 min at 20,000×*g* to remove the buffer solution.
14. The DNeasy Mini spin column is transferred to a 1.5-mL tube, and 100 μL of buffer AE is directly added onto the column membrane, incubated for 5 min at RT, and centrifuged for 1 min at 6,000×*g* to elute DNA.
15. Elution is repeated with another 100 μL buffer AE in the same tube.
16. One microliter of DNA solution is used to measure the concentration in a spectrophotometer and the rest is stored at –30°C.

3.2. Restriction Digestion of Genomic DNA

1. The solution containing 400 ng genomic DNA (from **Section 3.1**, step 15) is used for restriction digestion with 3 U of restriction digestion that does not have a recognition site on the gene of interest (*see* **Notes** 7 and **8**).
2. Digestion reaction is set with appropriate buffer in a total reaction volume of 30 μL overnight at 37°C.

3. The total volume of the reaction is adjusted to 400 μL by adding H_2O and an equal volume of phenol–chloroform–isoamyl alcohol (25:24:1) is added. The mixture is vortexed and centrifuged at 15,000×*g* for 10 min.
4. The supernatant is transferred to a new tube; 50 μL of 3 M sodium acetate and 1 mL of cold absolute alcohol are added to the tube and kept at –80°C for 1 h.
5. Solution is centrifuged at maximum speed (21,500×*g*) for 15 min at 4°C and the pellet washed twice with cold 70% ethanol.
6. Pellet is air-dried and resuspended in 20 μL H_2O and transferred to a 200-μL PCR tube.

3.3. Bisulfite Conversion of Genomic DNA

Restriction-digested genomic DNA is subjected to bisulfite conversion using the EZ DNA Methylation-Gold Kit (Zymo Research) according to the manufacturer's recommended procedure with slight modification:

1. The CT conversion reagent is prepared as follows: 900 μL H_2O, 300 μL M-dilution buffer, and 50 μL M-dissolving buffer are added to one tube of CT conversion reagent and vortexed for 10–15 min (*see* **Notes 9** and **10**).
2. An aliquot of 130 μL of freshly prepared CT conversion reagent is added to each DNA samples in a PCR tube (in **Section 3.2**, step 6). It is then mixed, centrifuged, and placed in a thermal cycler with the following program (*see* **Note 11**):
   a. 98°C for 10 min
   b. 53°C for 30 min
   c. Eight cycles of 53°C for 6 min and 37°C for 30 min
   d. Hold at 4°C
3. The Zymo-spin IC column is placed in a collection tube, and 600 μL of M-binding buffer is added the column.
4. The bisulfite-treated DNA samples (from **Section 3.3**, step 2) are added to the column.
5. Solution is centrifuged at maximum speed (21,500×*g*) for 30 s and the flow-through is removed.
6. One hundred microliters of M-wash buffer is added to the column and it is centrifuged for 30 s at a maximum speed.
7. After 200 μL of M-desulfonation buffer is added to the column, the column is kept for 20 min at RT (20–30°C) and centrifuged at maximum speed for 30 s.
8. The column is washed twice with 200 μL of M-wash buffer and placed in a 1.5-mL microcentrifuge tube.

9. Ten microliters of M-elution buffer is added directly to the column matrix, kept for 1 min, and the DNA eluted by centrifugation at maximum speed for 1 min.
10. Step 9 is repeated. Eluted DNA can be used for PCR amplification or stored at –80°C until use (*see* **Note 12**).

3.4. PCR Amplification of Bisulfite-Treated DNA and Gel Cleanup

1. PCR reaction is set in thin-walled tubes according to the following (*see* **Notes 13** and **14**): 1 μL of the bisulfite-treated DNA (300–500 pg) was used as a template for the PCR in a total volume of 50 μL containing 1 μM each PCR primer, 200 μM each dNTP, and 2.5 U of JumpStart *Taq* DNA polymerase in 10 mM Tris–HCl, pH 8.3, 50 mM KCl, and 3 mM $MgCl_2$.
2. PCR tubes were set in a thermal cycler with the following program (*see* **Note 15**):
   a. 94°C for 2 min
   b. 40 cycles of 94°C for 30 s, 45–60°C (depending on primer-Tm) for 30 s, 72°C for 45 s
   c. 72°C for 10 min
   d. Hold at 4°C
3. A low-melting agarose gel with 0.5 μg/mL ethidium bromide is prepared in 1× TAE buffer on a standard agarose gel electrophoresis system.
4. PCR reaction is mixed with DNA loading buffer and run on the gel at 80 V. DNA ladder of 100 bp may be used for size selection of PCR products.
5. Gel is visualized on a UV transilluminator and a DNA band of expected size is excised with a sterilized blade (*see* **Note 16**) and purified from the gel with QIAquick Gel Extraction Kit (Qiagen) (*see* **Note 17**).

3.5. Cloning and Screening of Amplified DNA Sequences

1. DNA purified from the specific gel fragment is used for ligation using the pGEM-T Easy Vector System according to the following: a total volume of 10 μL reaction mixture containing 5 μL of 2× rapid ligation buffer, 1 μL of 50 ng/μL pGEM-T Easy vector, 3 μL of gel-purified PCR product, and 1 μL of T4 DNA ligase.
2. Ligation reaction is mixed with pipetting, incubated at RT for 1 h, and kept at 4°C overnight; 2 μL of ligation reaction is used for transformation into 50 μL of JM109 high-efficiency competent cells according to the following:
   a. Aliquots of 50 μL of competent cells are transferred to a cold 1.5-mL tube to which 2 μL ligation reaction is added and mixed by flicking, then kept on ice for 30 min.

b. Cells are heat shocked at 42°C for 45 s and kept on ice for 2 min.

c. Cells are suspended in a total volume of 1 mL of SOC medium (preheated to 37°C) and shaken at 150–200 rpm for 60 min at 37°C.

d. The tubes are centrifuged at 700×*g* for 2 min and the bacterial pellet is resuspended in 200 μL SOC medium. The 50 μL of solution is spread on LB plates containing ampicillin/X-Gal/ IPTG and incubated at 37°C

A

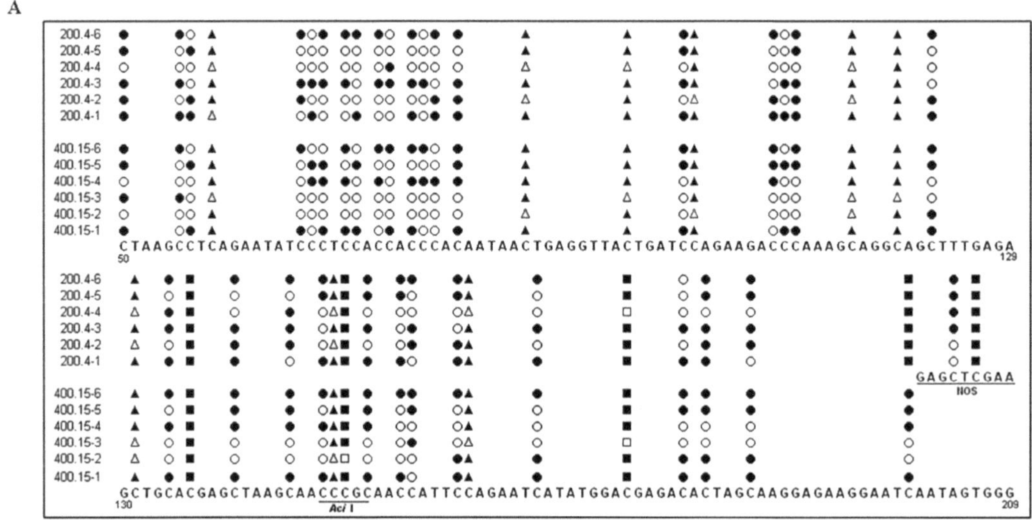

B

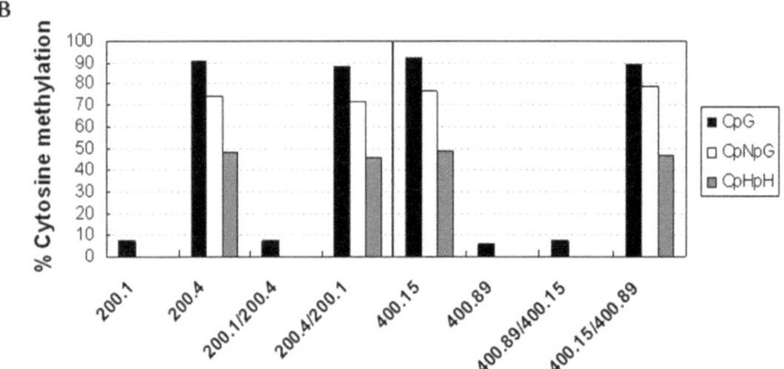

Fig. 13.1. Bisulfite sequence analysis of transgenic plants carrying 5′ region of *CP* gene of SPFMV. **a** The distribution of cytosine methylation is partially presented from the 50- to 206-bp region of CP. PCR amplification products from a bisulfite-treated DNA template were subcloned and then six independent clones were sequenced for each plant line (200.4-1 to 200.4-6 or 400.15-1 to 400.15-6). The locations of restriction sites for *AciI* are also shown. *Squares*, *triangles*, and *circles* indicate cytosine residues in CpG, CpNpG, and CpHpH contexts, respectively. *Filled* and *open symbols* indicate methylated and non-methylated cytosine residues, respectively. **b** A percent cytosine methylation in symmetrical (CpG and CpNpG) and non-symmetrical (CpHpH) contexts in the transgene coding region. Genomic DNA from silenced (200.4 and 400.15) and nonsilenced lines (200.1 and 400.89), or from non-silenced scions 6 weeks after grafting onto silenced rootstocks (200.1/200.4 and 400.89/400.15) or from silenced scions 6 weeks after grafting onto non-silenced rootstocks (200.4/200.1 and 400.15/400.89) was subjected to bisulfite sequencing. Percent cytosine methylation was analyzed for either the 5′ 200-bp region (for lines 200.1 and 200.4) or the 5′ 400-bp region of CP (for lines 400.15 and 400.89) and calculated from total cytosine residues present in the clones. (Reprinted from (6) with permission from Elsevier.)

overnight (~16 h). Generally white colonies indicate successful insertion of DNA into a plasmid vector.

3. Plasmid DNAs are isolated from 10 to 12 white colonies using the QIAprep Spin Miniprep Kit and the presence of insert is checked by digestion with *Eco*RI.

3.6. Sequencing of Clones

The plasmid DNA carrying insert (250–350 ng) is used for labeling for sequencing using the Bigdye Terminal Cycle Sequencing Kit (Applied Biosystems, Foster City, CA, USA) and subjected to sequencing. Sequencing reaction followed by sequence determination is performed according to the manufacturer's protocol (Applied Biosystems, Foster City, CA, USA).

3.7. DNA Methylation Analysis

The DNA sequence obtained is aligned with the untreated DNA sequence. Conservation of cytocines indicates the presence of methylation at those positions, while the conversion of cytocines to thymines indicates the absence of methylation (**Fig. 13.1**).

4. Notes

1. Do not mix buffer AP1 and RNase A before use: RNase A would be inactivated.
2. It is important to disrupt the tissue completely; presence of tissue clumps results in a lower yield of DNA.
3. Some plant materials can generate very viscous lysate and large amounts of precipitates resulting in shearing of DNA. This centrifugation step removes majority of the precipitates.
4. Use 1-mL tips to transfer the supernatant; otherwise, it might be necessary to cut off the end of tips.
5. A small amount of precipitate might cross the column and form a pellet in the collection tube, which must not be disturbed during the transfer.
6. Buffer AP3/E comes concentrated, so ethanol must be added prior to use.
7. We find that the restriction digestion of genomic DNA gives better reproducible results, though the DNA methylation kit used in this protocol does not require such digestion.
8. For example, we digested 400 ng of genomic DNA from transgenic plants with 3 U of *Eco*RV that does not have any recognition site on our gene of interest, *CP* of sweet potato feathery mottle virus (SPFMV).

9. CT conversion reagent is light sensitive; care should be taken to minimize its exposure to light. We wrap the tube with aluminum foil.
10. It is highly recommended to use freshly prepared CT conversion reagents in the DNA methylation kit. As each tube of CT conversion reagent is designed for 10 DNA samples, it is wise to treat all DNA samples at the same time to get consistent results.
11. This program gives better results to PCR amplify 400–500-bp DNA fragments. However, DNA template with extreme GC content (>80%) might result in less-efficient bisulfite conversion. An alternative program can be used in such a situation.
12. Bisulfite-treated DNA can be stored at or below –20°C. However, the conversion of cytocines to uracils by bisulfite treatment might lead DNA to fragile single-stranded conformation at this stage; thus it is highly recommended to store the samples at or below –80°C to prevent any degradation.
13. Optimization of the amount of DNA for PCR might be needed in some cases; starting with 0.5 μL up to 5 μL can be used in a 50-μL PCR reaction. We find that 1.0 μL is often sufficient for successful PCR.
14. Primer designing is most critical for bisulfite PCR. As a plant genome has DNA methylation on cytocines in any sequence context, degenerate nucleotides (Y for C/T in forward primer and R for G/A in reverse primer) are used to amplify any methylation state of cytocines. In our experience, 28–30-nt-long primers reduce non-specific PCR products. In addition, a couple of primers can be designed. It is recommended to design primers with Tm within 50–60°C. For example, 35S.BSP.F, 5′-ATATAA GGAAGTTYATTTYATTTGGAGAG-3′ and NOS.BSP.R, 5′-AACTTTATTRCCAAATRTTTRAACRAT-3′ primers were used to amplify bisulfite-treated DNA of 200- or 400-bp transgene sequences (6).
15. Optimization of PCR conditions might be required for different template DNA. Different annealing temperature, cycle number, and concentrations of Mg^{2+} can be verified in this situation.
16. Non-specific DNA fragments can be generated when a lower Tm is used for bisulfite PCR. If such non-specific DNA amplification cannot be avoided by optimizing the PCR reaction, a DNA fragment of expected size can be excised carefully and compared with the DNA ladder pattern of appropriate size marker.

17. Excised agarose gel with the expected DNA band can be directly used for ligation without purification with gel extraction kit; however, it often has lower ligation efficiency.

Acknowledgments

The authors wish to thank Dr. D Murphy for his English correction. This work was partially supported by the Ministry of Education, Culture, Sports, Science and Technology of Japan, Grant-in-Aid for Scientific Research, 00465001, 2009 and the Ministry of Agriculture, Forestry and Fisheries of Japan.

References

1. Ingelbrecht, I., van Houdt, H., Montagu, M. V., and Depicker, A. (1994) Posttranscriptional silencing of reporter transgenes in tobacco correlates with DNA methylation. *Proc. Natl. Acad. Sci. USA* **91**, 10502–10506.
2. Shapiro, R., Servis, R. E., and Welcher, M. (1970) Reactions of uracil and cytosine derivatives with sodium bisulfite: a specific deamination method. *J. Am. Chem. Soc.* **92**, 422–424.
3. Hayatsu, H., Wataya, Y., and Kai, K. (1970) The addition of sodium bisulfite to uracil and to cytosine. *J. Am. Chem. Soc.* **92**, 724–726.
4. Hayatsu, H., Wataya, Y., Kai, K., and Iida, S. (1970) Reaction of sodium bisulfite with uracil, cytosine, and their derivatives. *Biochemistry* **9**, 2858–2865.
5. Haque, A. K. M. N., Tanaka, Y., Sonoda, S., and Nishiguchi, M. (2007) Analysis of transitive RNA silencing after grafting in transgenic plants with the coat protein gene of sweet potato feathery mottle virus. *Plant Mol. Biol.* **63**, 35–47.
6. Haque, A. K. M. N., Yamaoka, N., and Nishiguchi, M. (2007) Cytosine methylation is associated with RNA silencing in silenced plats but not with systemic and transitive RNA silencing through grafting. *Gene* **396**, 321–331.
7. Frommer, M., McDonald, L. E., Millar, D. S., Collis, C. M., Watt, F., Grigg, G. W., Molloy, P. L., and Paul, C. L. (1992) A genomic sequencing protocol that yields a positive display of 5-methylcytosine residues in individual DNA strands. *Proc. Natl. Acad. Sci. USA* **89**, 1827–1831.
8. Gruna, C., Clark, S. J., and Rosenthal, A. (2001) Bisulfite genome sequencing: systematic investigation of critical experimental parameters. *Nucleic Acids Res.* **29**, e65.

Chapter 14

Using Nuclear Run-On Transcription Assays in RNAi Studies

Basel Khraiwesh

Abstract

RNA interference (RNAi) is a mechanism regulating gene transcript levels either by transcriptional gene silencing or by posttranscriptional gene silencing, which act in the genome maintenance and the regulation of gene expression which is typically inferred from measuring transcript abundance. Nuclear "run-on" (or "run-off") transcription assays have been used to obtain quantitative information about the relative rates of transcription of different genes in nuclei isolated from a particular tissue or organ. Basically, these assays exploit the activity of RNA polymerases to synthesize radiolabeled transcripts that then can be hybridized to filter-bound, cold, excess single-stranded DNA probes representing genes of interest. The protocol presented here streamlines, adapts, and optimizes nuclear run-on transcription assays for use in RNAi studies.

Key words: Gene expression, nuclear run-on, nuclei isolation, plants, RNAi, transcription.

1. Introduction

Since the discovery of RNAi in *Caenorhabditis elegans* (1), extensive studies have been performed focusing on the different aspects of RNAi. In particular, the elucidation of the essential components of RNAi pathways has advanced extensively (2). RNAi has been discovered in a wide range of organisms from plants and fungi to insects and mammals, suggesting that it arose early in the evolution of multicellular organisms (3). RNAi is a mechanism regulating gene transcript levels either by transcriptional gene silencing (TGS) or by posttranscriptional gene silencing (PTGS) (4). The RNAi pathway is typically initiated by ribonuclease III-like nuclease enzymes, called Dicer (DCL), that cleave double-stranded RNA molecules (dsRNAs; typically >200 nt) into small

H. Kodama, A. Komamine (eds.), *RNAi and Plant Gene Function Analysis*, Methods in Molecular Biology 744,
DOI 10.1007/978-1-61779-123-9_14, © Springer Science+Business Media, LLC 2011

fragments bearing a 3′ overhang of two nucleotides (5). One of these two strands is coupled to a second endonuclease enzyme called Argonaute (AGO) and then integrated into a large complex called RNA-induced silencing complex (RISC). It was proposed that small RNAs guide the cleavage of mRNA. These small RNAs regulate various biological processes, often by interfering with mRNA translation (2–4). Based on their biogenesis and function, small RNAs are classified as repeat-associated small interfering RNAs (ra-siRNAs), trans-acting siRNAs (ta-siRNAs), natural antisense transcript-derived siRNAs (nat-siRNAs), and microRNAs (miRNAs) (6). Recently, miRNAs have been identified as important regulators of gene expression in both plants and animals, and they are highly conserved in evolution (7–9). The mode of action of small RNAs to control gene expression at the transcriptional and posttranscriptional levels is now being developed into tools for biological research (10).

Gene expression is typically inferred from measuring transcript abundance. The overall process of gene expression can be considered to include all the factors that influence how much of a particular gene product is present at a particular time in an organism's development, under the influence of endogenous and exogenous effects. The process includes many steps, including signal transduction, transcriptional control, posttranscriptional effects on mRNA level, translational control, and protein turnover (11). RNA gel blotting (northern blotting), RT-PCR, and real-time quantitative PCR methods allow researchers to investigate the steady-state accumulation of a given RNA transcript in response to a given treatment or within a specific developmental state (12, 13). Microarrays extend this capacity to the entire transcriptome in one experimental trial (14, 15). All methods mentioned above provide a means to report the relative steady-state abundance of the molecule or molecules in question a snapshot of the state of the mRNA species of interest at a given point in time. Methods to assay transcription rate directly have allowed researchers to reconcile discrepancies between presumed transcription rates and transcript accumulation.

Newly transcribed RNA can be identified using the nuclear "run-on" (or "run-off") transcription assays (16). The first step of the nuclear run-on assay is the isolation of intact nuclei, because newly synthesized mRNA can be labeled with high specific activity in isolated nuclei compared to intact cells. The nuclei are incubated in the presence of radiolabeled UTP allowing the in vitro transcription of a labeled total mRNA over a defined time. After the transcription reaction, the radiolabeled RNA is then isolated and hybridized with denatured DNA, which has been immobilized on a filter membrane. This allows the detection of specific mRNA transcripts. The DNAs usually include cDNA sequences of the gene(s) of interest and of a housekeeping gene that is stably expressed as a standard. Isolation of transcriptionally active nuclei and their subsequent use in nuclear run-on assays has been performed in many plant species including pea, tomato, rice, petunia, soybean, *Arabidopsis*, and *Physcomitrella* (10, 17–22). Nuclear run-on assays and detection of gene transcription activity from isolated nuclei showed miRNA-directed transcriptional silencing in mammalian cells (23, 24). Recently, nuclear run-on assays in *Physcomitrella patens* showed that hypermethylation of miRNA target loci leads to transcriptional silencing of miRNA target genes (10). These examples illustrate how nuclear run-on assays provide important information related to RNAi studies. Here we report the optimization of nuclear run-on assays for RNAi studies in plants.

2. Materials

Use autoclaved pipette tips and baked glassware. Wear gloves when you work. In most lab manuals it is recommended to prepare all required solutions with diethylpyrocarbonate (DEPC)-treated H_2O.

2.1. Plant Materials

1. Culture media and growth conditions depend on plant species.
2. Fresh tissue.

2.2. Filter-Bound Probe Assembly

1. cDNA plasmid (representing genes of interest).
2. 0.4 M NaOH.
3. 20X SSC buffer: 3.0 M NaCl and 0.3 M sodium citrate, pH 7.0.
4. Nylon blotting membrane (0.45 μm nitrocellulose membrane).
5. Slot blot apparatus.
6. Vacuum oven.
7. UV cross-linker (Biolink BLX, Biometra or equivalent).

2.3. Isolating Nuclei from Tissue

1. Nuclei extraction buffer: 2.0 M hexylene glycol (2-methyl-2,4-pentanediol), 20 mM PIPES-KOH, pH 7.0, 10 mM $MgCl_2$, and 5 mM 2-mercaptoethanol. Prepare as 5X and store without 2-mercaptoethanol at 4°C. Add 2-mercaptoethanol immediately before use (*see* **Note 1**).
2. Filter mesh 100 or cheesecloth (coarse cotton gauze).
3. Triton X-100 (final concentration of Triton X-100 depends on plant species, *see* **Note 2**).

4. Percoll.
5. Gradient buffer (1X): 0.5 M hexylene glycol, 5 mM PIPES-KOH, pH 7.0, 10 mM $MgCl_2$, 5 mM 2-mercaptoethanol, and 1% Triton X-100. Prepare as 5X and store without 2-mercaptoethanol at 4°C. Add 2-mercaptoethanol immediately before use (*see* **Note 1**).
6. Nuclei storage buffer: 50 mM Tris–HCl, pH 7.8, 10 mM 2-mercaptoethanol, 20% glycerol, 5 mM $MgCl_2$, and 0.44 M sucrose (*see* **Note 1**).
7. Trypan blue.
8. 10 mM dithiothreitol (DTT).
9. Pasteur pipettes and appropriate bulbs.
10. 25 mL glass or plastic pipettes with appropriate pump.
11. Test tubes (50, 14, 1.5 mL).
12. Centrifuge (Sorvall with SS-34 rotor or equivalent).
13. Microcentrifuge (Eppendorf 5417R or equivalent).
14. Liquid nitrogen.
15. Mortar and pestle.
16. Funnel.
17. Light microscopy.
18. DAPI staining buffer and microscope (optional).

2.4. Radiolabeling of the Transcripts in Nuclei Isolated from Tissue

1. Placental ribonuclease inhibitor (RNasin, Promega).
2. Transcription assay buffer (10X): 250 mM Tris–HCl, pH 7.8, 375 mM NH_4Cl, 50 mM $MgCl_2$, and 50% (v/v) glycerol. Store at <0°C and use at 1X.
3. 100 mM stocks of CTP, GTP, and ATP (Epicenter Inc.).
4. Uridine 5′-[α-^{32}P]triphosphate, triethylammonium salt (3,000 Ci $mmol^{-1}$). Use 100 μCi for each sample of nuclei to be radiolabeled.
5. RNase-free water.
6. DNase I.
7. Termination buffer: 7.5 M urea, 0.5% SDS, 20 mM EDTA, pH 7.5, and 100 mM LiCl.
8. Phenol:chloroform:octanol reagent, 25:24:1, pH 7.6. To pH phenol solutions with a pH meter, mix 2 mL of the organic phase with 8 mL of methanol and 10 mL of H_2O. Adjust the pH using phenol equilibration buffer (AppliChem GmbH, Germany).
9. 7.5 M ammonium acetate, pH 7.0.
10. 200 mg mL^{-1} yeast tRNA.

11. 100% ethanol.
12. 20X SSC buffer: 3.0 M NaCl and 0.3 M sodium citrate, pH 7.0.
13. Prehybridization/hybridization buffer (varies with laboratory preference): 50% (v/v) formamide, 6X SSC, 5 mM sodium pyrophosphate, 2X Denhardt's solution, 0.5% (w/v) SDS, 10 μg/mL poly(A), and 100 μg/mL salmon sperm DNA.
14. Washing solution: 1X SSC, 0.1% SDS.
15. PhosphorImager (Imager FX, Bio-Rad or equivalent).
16. Microcentrifuge, dedicated for radioisotope use.
17. Geiger counter.
18. Plastic wrap.
19. Hybridization rotary oven.
20. Water bath.

3. Methods

Experiments involving RNA require precautions to prevent RNA degradation.

3.1. Filter-Bound Probe Assembly

1. Assemble slot blot apparatus, prewet each slot with 10X SSC buffer.
2. Denature 1 μg of each DNA probe (*see* **Note 3**) in separate microfuge tubes in 50 μL 0.4 M NaOH at 65°C for 10 min.
3. Remove from heat, transfer to ice, and add 200 μL 20X SSC.
4. Apply the entire volume of each denatured probe to a nylon blotting filter via a slot blot manifold and apply a gentle vacuum to attach cold DNA probes to the nylon filter.
5. Rinse the slot blot wells with 10X SSC.
6. Including a small amount of DNA gel loading dye allows visualization of the probe on the filter for ease in labeling.
7. Crosslink the probes to filter at optimal setting using UV cross-linker (Biolink BLX, Biometra or equivalent).
8. Store the filter until use.

3.2. Isolating Nuclei from Tissue

Perform all steps at 2–8°C. Use pre-cooled buffers and equipment. Be sure that all the solutions are mixed thoroughly. All centrifugations should be performed at 2–8°C with pre-cooled rotors.

1. Grind 2–4 g fresh plant tissue (*see* **Note 4**) with liquid nitrogen to a fine powder using a mortar and pestle. Let the

liquid nitrogen evaporate completely. The color changes from gray-white to dark-green. It is recommended to transfer the powder from the cold mortar to a new container before addition of the buffer to prevent the buffer from freezing.

2. Resuspend the powder in 20–40 mL (~10 mL/g tissue) of ice-cold extraction buffer and homogenize gently using a Polytron homogenizer (or comparable device) set on its lowest speed (approximately for 8–10 min).
3. Soak three to five layers of cheesecloth with extraction buffer and place over the top of a 50 mL centrifuge tube or 50 mL beaker. Slowly decant the homogenate through the cheesecloth filter and, upon completion, squeeze the cloth to remove excess plant extract into the beaker.
4. Add extraction buffer to the extract in the tube or beaker to a final volume of 20–40 mL (~10 mL/g tissue).
5. Gently stir the resulting extract on ice using a magnetic stir plate. Slowly add 25% Triton X-100 dropwise to the solution until the final concentration is 1% (*see* **Notes 2** and **5**).
6. Prepare 80 and 30% Percoll suspensions by mixing 100% Percoll with 5X gradient buffer and water to achieve a final concentration of 1X gradient buffer. Prepare 6 mL of 80% and 12 mL of 30% per sample.
7. Add 6 mL of the 30% Percoll solution in 1X gradient buffer into a 50 mL round-bottomed centrifuge tube. Using a Pasteur pipette or 5 mL pipettor, gently add 6 mL of the 80% Percoll solution to the bottom of the test tube by passing the pipette through the 30% layer and slowly expelling the solution into the very bottom of the tube (*see* **Note 6**).
8. Carefully apply the filtered plant extract onto the top of the 30% Percoll layer (30 mL extract in standard preparation, 1.5 mL into two-step gradients for the small-scale isolation). This is most easily performed by drawing the extract into a Pasteur pipette and slowly releasing the solution onto the side of the test tube directly above the 30% layer (**Fig. 14.1a**).
9. Centrifuge the gradient at 2,000×*g* for 30 min at 4°C. It is best to use a swinging bucket rotor; however, fixed-angle rotors may be used. After centrifugation, the nuclei will be present as a conspicuous dark band at the interface between the 30 and 80% fractions (**Fig. 14.1b**).
10. Gently collect the nuclei band; draw off the top layer of extraction buffer using a 25 mL glass pipette, taking extreme care not to allow inadvertent backflow that could

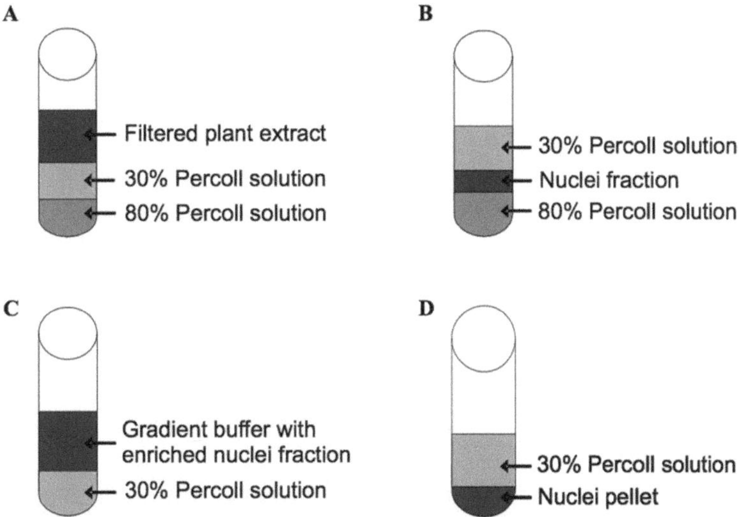

Fig. 14.1. Fractionation by Percoll gradients. (**a**) An example of the step gradients before centrifugation (*see* **Section 3.2**, Steps 6–8). (**b**) An example of the step gradients after centrifugation. The location of the 30 and 80% Percoll gradients is noted and the nuclei fraction present as a conspicuous dark band at the interface between the 30 and 80% fractions (*see* **Section 3.2**, Steps 9 and 10). (**c**) Accumulation of the enriched nuclei fraction is shown at the 30% Percoll interface (*see* **Section 3.2**, Step 11). (**d**) Accumulation of the nuclei pellet after centrifugation (*see* **Section 3.2**, Step 12).

disrupt the gradient. Use a Pasteur pipette to collect nuclei from the 30 to 80% interface. The nuclei will tend to travel together as small sticky clumps. It is important to minimize the amount of 80% Percoll removed with the nuclei fraction. Move nuclei to a clean, cold 50 mL test tube (*see* **Note 6**).

11. Add 1X gradient buffer to the enriched nuclei fraction to a total volume of 10 mL and then underlay the solution with 6 mL of 30% Percoll. Again, this is achieved by placing a Pasteur pipette filled with 30% Percoll through the nuclei/gradient buffer solution to the bottom of the test tube and gently expelling the buffered Percoll solution (**Fig. 14.1c**).
12. Centrifuge at 2,000×*g* for 10 min at 4°C and gently decant the supernatant, containing gradient buffer and Percoll. Upon close inspection, the pellet will likely contain multiple layers. Typically, these are off-white (starch), green (other debris), and gray, from bottom to top. The nuclei reside in the top, gray layer (**Fig. 14.1d**).
13. Resuspend the resulting nuclei pellet in 1X gradient buffer (about 1.5 mL) and transfer to a microcentrifuge tube. Centrifuge gently (<1,000×*g*) to pellet the nuclei. This pellet is gently resuspended in approximately three volumes of nuclei storage buffer ~500 μL (*see* **Note 7**).

14. At this point, the nuclei may be diluted and quantified using a hemocytometer and either used directly or stored at –70°C in aliquots of 10^6 nuclei. Nuclei may be visualized using DAPI staining if the appropriate microscope and filters are available. Simple quantification is achieved by using light microscopy. The nuclei are readily visualized by this method when stained with trypan blue.

3.3. Radiolabeling of the Transcripts in Nuclei Isolated from Tissue

1. Start with approximately 10^6 nuclei in 50 μL of nuclei storage buffer (*see* **Section 3.2**, Step 13). Starting with an approximately equal number of nuclei will provide better results.
2. Add 20 U RNasin to nuclei and incubate at 30°C in a water bath for 10 min.
3. Set up the run-on assay as detailed below. The first four components should be combined and prewarmed to 30°C before addition to the nuclei:

10X transcription assay buffer	10 μL
100 mM CTP, GTP, and ATP	5 μL each
[α-^{32}P]UTP	10 μL
RNase-free water	15 μL
Nuclei in storage buffer	50 μL

4. Allow the reaction to proceed at 30°C for 30 min.
5. Terminate the reaction by adding 10 U of DNase I and incubating at 30°C for 10 min.
6. Add 200 μL of termination buffer and mix thoroughly.
7. Add 300 μL of phenol:chloroform:octonol reagent and vortex for 1 min.
8. Centrifuge for 5 min in microcentrifuge at full speed (typically 20,000×*g*, or ~12,000 rpm. in a standard centrifuge).
9. Transfer aqueous fraction (top fraction) to a clean microcentrifuge tube and add 1/10 volume 7.5 M ammonium acetate, pH 7.0, and 100 μg of yeast tRNA before adding two volumes of ice-cold ethanol. Place into a dry ice/ethanol bath for 10 min.
10. Centrifuge for 10 min at full speed in a microcentrifuge. Discard the supernatant. Resuspend the pellet in 500 μL hybridization buffer and heat for 5 min at 95°C.
11. Pre-hybridize the filter containing the slot-blotted hybridization probes (*see* **Section 3.1**, Step 8) in hybridization buffer at 60°C for at least 1 h. Calculate the amount of buffer as 0.1 mL buffer per cm^2 of nylon filter minus 0.5 mL that will be added with the labeled transcripts.
12. Add the heat-denatured, in vitro-synthesized transcripts from Step 10 to the hybridization vessel and incubate in a

hybridization oven, water bath, or other suitable treatment that will ensure constant motion of buffer for at least 12 h. Hybridize at 60°C, the hybridization may run as long as 24 h (*see* **Note 8**).

13. Following hybridization, remove the blot from the bag or bottle and wash it in three consecutive solutions of 1X SSC/0.1% SDS, 10 min each at 60°C (*see* **Notes 8** and **9**).
14. Quantify hybridization signals by autoradiography or phosphorimaging. If signal is detected in the negative control lane, repeat Steps 13 and 14 using more stringent washes, until the nonspecific controls are no longer detectable (*see* **Note 10**). An example of experimental results is shown in **Fig. 14.2**.

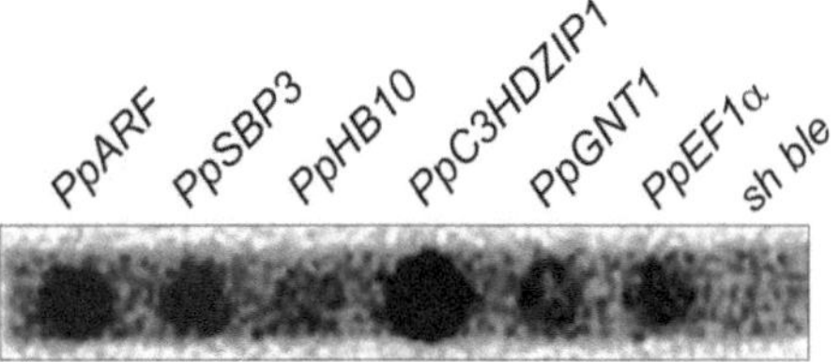

Fig. 14.2. Nuclear run-on assay from *Physcomitrella patens*. Membrane with dot-blotted cDNA fragments of the miRNA target genes *PpARF*, *PpSBP3*, *PpHB10*, and *PpC3HDZIP1*; the control genes *PpGNT1* and *PpEF1α*; and the negative control bacterial zeocin resistance gene *Sh ble* (*see* **Note 11**). The membrane was hybridized with radio-labeled RNA obtained by nuclear run-on assay from *P. patens* wild-type plant. Hybridization signals were normalized to the control gene *PpEF1α* (10).

4. Notes

1. Preparation of buffers and reagents is time consuming and therefore all concentrated buffers, 25% (v/v) Triton, and nuclear run-on buffers should be prepared at least a day before initiating the nuclei isolation. Although many of the reagents may be made ahead of time, it is important to leave out 2-mercaptoethanol until immediately before use.
2. Each type of plant requires a different percentage of Triton X-100 for proper lysis. Take several microliters of the cells in extraction buffer and view them under the microscope. If a large proportion of cells remain intact, add more Triton X-100, until most of the cells have lysed. If lysis of nuclei or a gelatinous mass is observed, lyse the cells with a lower final concentration of Triton X-100.
3. Probes typically consist of PCR amplicons or plasmid inserts that have been gel purified and quantified beforehand. The optimal probes are amplified from cDNA and represent the full coding region of the gene. One microgram is required per experimental condition. It

is important to include several standards and controls. Typically genes considered to be constitutively expressed make useful standards that allow comparison between samples.

4. For better yield of nuclei and nuclear RNA, it is recommended to use relatively young fresh leaves. Different tissues of the same plant can give different yields of nuclei and nuclear RNA.
5. It is important to avoid creating local high concentrations of Triton X-100. The 25% solution should be added as slowly as possible to a moving plant extract, gently stirring on a magnetic stir plate. It is best to use a pipettor or Pasteur pipette where microliter amounts can be continually introduced to the solution.
6. Pipetting the 80% fraction under the 30% fraction produces a more defined interface when compared to placing the 30% on top of the 80% layer.
7. Nuclei may be stored at –70°C and remain transcriptionally active for at least 6 months.
8. Hybridization conditions may differ substantially based on the intention of the designed experiment. The most important criteria are the stringency conditions. High hybridization and wash temperature coupled with low salt concentrations are high-stringency hybridization conditions. Low temperature and high salt concentrations favor unspecific hybridizations and represent low-stringency hybridization conditions.
9. All radioactive hybridization buffers and washes should be discarded in accordance with acceptable practices, which include national guidelines.
10. Hybridization signals should be visible within days of exposure, using standard autoradiography film and associated enhancer screens. If the signals are not visible, then it may be advisable to lengthen the exposure time to avoid premature development.
11. The transcription rate of any given gene can only be estimated when compared to internal controls. Therefore, the power of an experiment relies on the presence of several control probe sets to monitor genes known to be constitutively expressed.

Acknowledgments

The author would like to thank Dr. Enas Qudeimat and Dr. Mieke Van Lijsebettens for helpful comments on the manuscript.

References

1. Fire, A., Xu, S., Montgomery, M. K., Kostas, S. A., Driver, S. E., and Mello, C. C. (1998) Potent and specific genetic interference by double-stranded RNA in *Caenorhabditis elegans*. *Nature* **391**, 806–811.
2. Tomari, Y., and Zamore, P. D. (2005) Perspective: Machines for RNAi. *Genes Dev.* **19**, 517–529.
3. Hannon, G. J. (2002) RNA interference. *Nature* **418**, 244–251.
4. Meister, G., and Tuschl, T. (2004) Mechanisms of gene silencing by double-stranded RNA. *Nature* **431**, 343–349.
5. Myers, J. W., Jones, J. T., Meyer, T., and Ferrell, J. E., Jr. (2003) Recombinant Dicer efficiently converts large dsRNAs into siRNAs suitable for gene silencing. *Nat. Biotechnol.* **21**, 324–328.
6. Chapman, E. J., and Carrington, J. C. (2007) Specialization and evolution of endogenous small RNA pathways. *Nat. Rev. Genet.* **8**, 884–896.
7. Axtell, M. J., and Bowman, J. L. (2008) Evolution of plant microRNAs and their targets. *Trends Plant Sci.* **13**, 343–349.
8. Khraiwesh, B., Ossowski, S., Weigel, D., Reski, R., and Frank, W. (2008) Specific gene silencing by artificial microRNAs in *Physcomitrella patens*: An alternative to targeted gene knockouts. *Plant Physiol.* **148**, 684–693.
9. Rajagopalan, R., Vaucheret, H., Trejo, J., and Bartel, D. P. (2006) A diverse and evolutionarily fluid set of microRNAs in *Arabidopsis thaliana*. *Genes Dev.* **20**, 3407–3425.
10. Khraiwesh, B., Arif, M. A., Seumel, G. I., Ossowski, S., Weigel, D., Reski, R., and Frank, W. (2010) Transcriptional control of gene expression by microRNAs. *Cell* **140**, 111–122.
11. Darnell, J. E., Jr. (1982) Variety in the level of gene control in eukaryotic cells. *Nature* **297**, 365–371.
12. Alwine, J. C., Kemp, D. J., and Stark, G. R. (1977) Method for detection of specific RNAs in agarose gels by transfer to diazobenzyloxymethyl-paper and hybridization with DNA probes. *Proc. Natl. Acad. Sci. USA* **74**, 5350–5354.
13. Higuchi, R., Fockler, C., Dollinger, G., and Watson, R. (1993) Kinetic PCR analysis: Real-time monitoring of DNA amplification reactions. *Biotechnology (NY)* **11**, 1026–1030.
14. Galbraith, D. W. (2006) DNA microarray analyses in higher plants. *OMICS* **10**, 455–473.
15. Gutierrez, R. A., Ewing, R. M., Cherry, J. M., and Green, P. J. (2002) Identification of unstable transcripts in Arabidopsis by cDNA microarray analysis: Rapid decay is associated with a group of touch- and specific clock-controlled genes. *Proc. Natl. Acad. Sci. USA* **99**, 11513–11518.
16. Greenberg, M. E., and Bender, T. P. (2007) Identification of newly transcribed RNA. *Curr. Protoc. Mol. Biol.* **4**, 10.1–10.7.
17. Chan, M. T., and Yu, S. M. (1998) The 3′ untranslated region of a rice α-amylase gene mediates sugar-dependent abundance of mRNA. *Plant J.* **15**, 685–695.
18. Dean, C., Favreau, M., Bond-Nutter, D., Bedbrook, J., and Dunsmuir, P. (1989) Sequences downstream of translation start regulate quantitative expression of two petunia *rbcS* genes. *Plant Cell* **1**, 201–208.
19. Folta, K. M., and Kaufman, L. S. (2006) Isolation of *Arabidopsis* nuclei and measurement of gene transcription rates using nuclear run-on assays. *Nat. Protoc.* **1**, 3094–3100.
20. Giuliano, G., and Scolnik, P. A. (1988) Transcription of two photosynthesis-associated nuclear gene families correlates with the presence of chloroplasts in leaves of the variegated tomato *ghost* mutant. *Plant Physiol.* **86**, 7–9.
21. Lescure, A. M., Proudhon, D., Pesey, H., Ragland, M., Theil, E. C., and Briat, J. F. (1991) Ferritin gene transcription is regulated by iron in soybean cell cultures. *Proc. Natl. Acad. Sci. USA* **88**, 8222–8226.
22. Marrs, K. A., and Kaufman, L. S. (1989) Blue-light regulation of transcription for nuclear genes in pea. *Proc. Natl. Acad. Sci. USA* **86**, 4492–4495.
23. Gonzalez, S., Pisano, D. G., and Serrano, M. (2008) Mechanistic principles of chromatin remodeling guided by siRNAs and miRNAs. *Cell Cycle* **7**, 2601–2608.
24. Kim, D. H., Saetrom, P., Snove, O., Jr., and Rossi, J. J. (2008) MicroRNA-directed transcriptional gene silencing in mammalian cells. *Proc. Natl. Acad. Sci. USA* **105**, 16230–16235.

Chapter 15

Proteomic Analysis of RNA Interference Induced Knockdown Plant

Sang Yeol Lee and Kyun Oh Lee

Abstract

RNA interference (RNAi) is a useful research tool for the specific deletion, or knockdown, of target genes that can be exploited both in cultured plant cells and in whole plants. In RNAi, hairpin RNA (hpRNA)-transduced lines are used to identify loss-of-function mutations in multi-copy genes with redundant functions in polyploid plant species. Plants transformed with hpRNA exhibit a range of phenotypes resulting from complete knockdown to weak suppression or tissue- and stage-specific knockdown. Functional genomics using proteomic analysis with two-dimensional gel electrophoresis (2-DE) and mass spectrometry provides valuable information about altered levels of expression of specific genes in biological samples. Here, we describe the proteomic analysis of *Oryza sativa* (*Os*) thioredoxin *m* (*Ostrxm*) knockdown using 2-DE and matrix-assisted laser desorption/ionization time-of-flight mass spectrometry (MALDI-TOF MS).

Key words: 2-DE, hpRNA, knockdown, MALDI-TOF MS, proteomics, RNAi.

1. Introduction

Thioredoxins are a ubiquitously distributed family of proteins that regulate a variety of cellular processes through the function of their highly conserved active site sequence, Trp-Cys-Gly-Pro-Cys (1). In contrast to prokaryotic and other eukaryotic cells, plants contain several thioredoxin variants (2–4). The *Arabidopsis* genome contains at least eight genes for thioredoxin *h*, two for thioredoxin *f*, and four for thioredoxin *m*, all of which localize to different tissues or intracellular organs (3, 5, 6). The function of chloroplast thioredoxins from cyanobacteria to higher plants has been implicated in light-dependent regulation of the activity of photosynthetic enzymes (7). In the light, reduction of specific

H. Kodama, A. Komamine (eds.), *RNAi and Plant Gene Function Analysis*, Methods in Molecular Biology 744,
DOI 10.1007/978-1-61779-123-9_15, © Springer Science+Business Media, LLC 2011

disulfide bridges by chloroplast thioredoxins results in activation of target proteins (8–10). In the dark, thioredoxins and their substrates are present in an oxidized state that is enzymatically inactive (11). Thioredoxins *f* and *m* can activate fructose bisphosphatase (FBPase) and NADP-dependent malate dehydrogenase (MDH) in vitro, which suggests that there is some overlap in the substrate specificity of the thioredoxin isoforms (12, 13). Importantly, disruption of the gene for thioredoxin *m* in cyanobacteria (*Anacystis nidulans*) is lethal, indicating that this protein may have critical and distinct functions in vivo. These results prompted us to investigate the functional roles of thioredoxins in intact organisms using a variety of genetic tools (14).

Although the molecular characteristics and biochemical properties of chloroplast thioredoxins have been well-studied in vitro (3, 5, 6, 11, 12, 15, 16), less is known about their physiological roles in intact plants. Many duplicated isoforms of small molecular weight (12–14 kDa) thioredoxins and/or subfamilies of these proteins are found in plants. Some T-DNA insertion mutants of the thioredoxins do not display any detectable phenotype, possibly due to functional redundancy among the various isoforms. Even the probability of random mutagenesis by T-DNA insertion is relatively low due to the small size of the corresponding genomic sequences. The technique of RNAi has been used to investigate or validate the functions of a number of duplicated genes. In the RNAi pathway, small interfering RNA (siRNA) is incorporated into an RNA-induced silencing complex (RISC), which selectively targets messenger RNA to prevent its translation (17). Here, we describe the generation of *Oryza sativa* thioredoxin *m* (*Ostrxm*) knockdown rice using RNAi and analysis of the expression of other proteins in *Ostrxm* knockdown plants by 2-DE and MALDI-TOF MS (18).

2. Materials

2.1. Plant and Growth Conditions

O. sativa L. cv. Dong-jin was used for all experiments. Plants were grown in a growth chamber conditioned with supplemental lighting (8 h of dark, 16 h of light; 200 μmol m^{-2} s^{-1}) and a day/night temperature cycle of 30/25°C or in a greenhouse with sunlight.

2.2. Media for Rice Transformation

1. YEP medium: 1% (w/v) yeast extract, 1% (w/v) peptone, and 0.5% (w/v) sodium chloride, pH 7.0 (*see* **Note 1**).
2. MSJ4 medium: 0.441% (w/v) basal salt mixture containing micro- and macro-elements with vitamins (Duchefa M0222) (19, 20), pH 5.8, 0.01% (w/v) myo-inositol, 0.1% (w/v) L-proline, 0.05% (w/v) L-alanine, 0.03% casein

enzymatic hydrolysate, 0.0002% (w/v) 2,4-D, 0.025% (w/v) cefotaxime, 0.005% (w/v) hygromycin, 3% (w/v) sucrose, and 0.3% (w/v) phytagel (*see* **Note 1**).

3. MSJ2 medium: 0.441% (w/v) basal salt mixture containing micro- and macro-elements with vitamins (19, 20), pH 5.8, 0.01% (w/v) myo-inositol, 0.1% (w/v) L-proline, 0.05% (w/v) L-alanine, 0.03% casein enzymatic hydrolysate, 0.0002% (w/v) 2,4-D, 0.025% (w/v) cefotaxime, 0.005% (w/v) hygromycin, 1.5% (w/v) maltose, 1% (w/v) glucose, and 0.3% (w/v) phytagel; acetosyringone was added at a final concentration of 200 μM after sterilization of the medium (*see* **Note 1**).
4. MSR medium: 0.441% (w/v) basal salt mixture containing micro- and macro-elements with vitamins (19, 20), pH 5.8, 0.0001% (w/v) NAA (auxin), 0.0005% (w/v) kinetin, 0.025% (w/v) cefotaxime, 0.005% (w/v) hygromycin, 3% (w/v) D-sorbitol, 1.5% (w/v) maltose, and 0.4% (w/v) phytagel (*see* **Note 1**).
5. 10X AA macronutrients: 2.95% (w/v) KCl, 0.25% (w/v) $MgSO_4 \cdot 7H_2O$, 0.15% (w/v) $CaCl_2 \cdot 2H_2O$, and 0.15% (w/v) $NaH_2PO_4 \cdot H_2O$ (*see* **Note 2**).
6. 1000X AA micronutrients: 0.1% (w/v) $MnSO_4 \cdot 7H_2O$, 0.03% (w/v) H_3BO_3, 0.02% (w/v) $ZnSO_4 \cdot 7H_2O$, 0.0075% (w/v) KI, 0.0025% (w/v) $Na_2MoO_4 \cdot 2H_2O$, 0.00025% (w/v) $CuSO_4 \cdot 5H_2O$, and 0.00025% (w/v) $CoCl_2 \cdot 6H_2O$ (*see* **Note 2**).
7. 100X $FeSO_4$ EDTA AA iron: 0.278% (w/v) $FeSO_4 \cdot 7H_2O$ and 0.373% (w/v) Na_2EDTA; stock solution was stored at 4°C in a dark bottle (*see* **Note 2**).
8. 100X AA amino acids: 8.76% (w/v) L-glutamine, 2.66% (w/v) L-aspartic acid, 1.74% (w/v) L-arginine, and 0.075% (w/v) glycine (*see* **Note 2**).
9. 100X MS vitamins: 1% (w/v) myo-inositol, 0.005% (w/v) nicotinic acid, 0.005% (w/v) pyridoxine HCl, 0.001% (w/v) thiamine HCl, and 0.002% (w/v) glycine (*see* **Note 2**).
10. AA medium: 1X AA macronutrients, 1X AA micronutrients, 1X $FeSO_4$ EDTA AA iron, 1X AA amino acids, 1X MS vitamins, 6.85% (w/v) sucrose, 3.6% (w/v) D-glucose, and 0.05% (w/v) casamino acids, pH 5.2, 200 μM acetosyringone (*see* **Notes 2** and **3**).

2.3. Southern and Northern Blot Analyses

1. RNA extraction buffer: 4 M guanidinium isothiocyanate, 50 mM Tris–HCl, 5 mM sodium citrate, pH 7.0, 10 mM EDTA, 5% β-mercaptoethanol, and 5% sarkosyl (*see* **Note 4**).

2. 10X RNA gel running buffer: 0.2 M MOPS, 50 mM sodium acetate, and 10 mM EDTA, pH 7.0.
3. RNA gel: 1.2% agarose, 1X RNA gel running buffer, and 7% formaldehyde.
4. 20X SSC: 3 M NaCl, 0.3 M sodium citrate, pH 7.3.

2.4. RT-PCR

1. PCR mixture: 100 ng cDNA as template, 5 pmol each oligonucleotide primer, 200 μm each dNTP, 1 unit Ex Taq polymerase (Takara Bio), and 1X Ex Taq polymerase buffer in a 20 μL volume.
2. Oligonucleotide primers: Os02g42700F, 5′-TCGCCAAGCCAGCCGCGTCTC-3′; Os02g42700R, 5′-CAGGGCTCAGCCTTGAATTGCATTTAT-3′; Os04g44830F, 5′-CCGTCTGCTCCTCTCCCCTCG-3′; Os04g44830R, 5′-CAGGCTTCCTTGCGACTTTCATGATTT-3′; Os12g08730F, 5′-ATGGCGTTGGAGACGTGCTTC-3′; Os12g08730R, 5′-GCCACAGTATTGATTCCATGGAGTAGTAGGAT-3′.

2.5. Materials and Buffer for 2-DE

1. Urea, thiourea, 3-[(3-cholamidopropyl)dimethylammonio]-1-propanesulfonate (CHAPS), dithiothreitol (DTT), benzamidine, Bradford solution, acrylamide, iodoacetamide, bisacrylamide, SDS, acetonitrile, trifluoroacetic acid, and α-cyano-4-hydroxycinnamic acid were purchased from Sigma-Aldrich (Electrophoresis grade, ACS reagents, ultrapure). Pharmalyte (pH 3.5–10) was from Amersham Biosciences, IPG DryStrips (pH 4–10 NL, 24 cm) were from Genomine Inc., and modified porcine trypsin (sequencing grade) was from Promega.
2. Sample lysis buffer: 7 M urea, 2 M thiourea containing 4% (w/v) CHAPS, 1% (w/v) DTT, 2% (v/v) pharmalyte, and 1 mM benzamidine.
3. Equilibration buffer: 6 M urea, 2% (w/v) SDS, 30% (v/v) glycerol, and 50 mM Tris–HCl, pH 6.8.

3. Methods

Here, we describe the use of hpRNA-induced RNAi to suppress the expression of *Ostrxm*. Using the procedures outlined below, we initially generated 15 hygromycin-resistant T_0 transformants, which were transferred to soil for acclimatization. Of the 15 hygromycin-resistant T_0 plants, only 4 lines survived and produced T_1 seeds. If an RNAi target gene is essential for plant development, complete or near-complete suppression of transcription

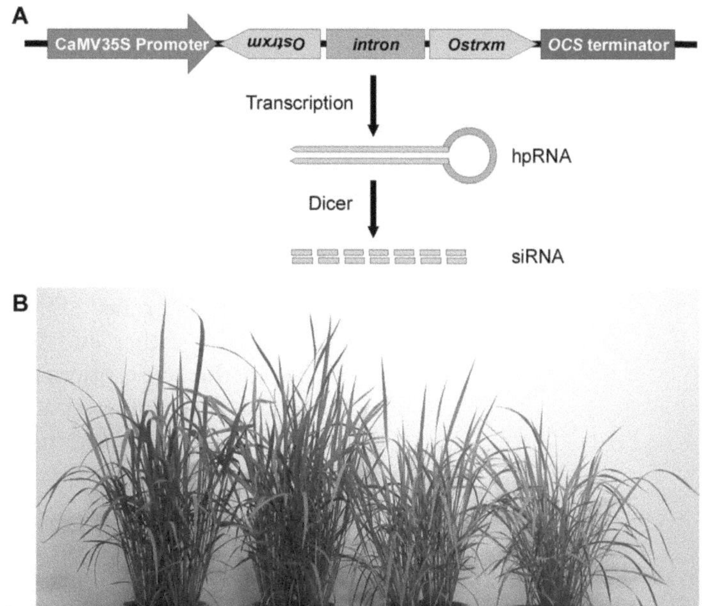

Fig. 15.1. The *Ostrxm* RNAi construct and phenotype of *Ostrxm* RNAi rice plants. (**a**) Schematic representation of the *Ostrxm* RNAi construct, hpRNA and siRNA. (**b**) Phenotype of WT and transgenic rice plants grown for 70 days. WT and transgenic rice plants were grown under sunlight conditions in a greenhouse. WT, vector control, *Ostrxm* RNAi #4 and #14 plants are arranged from *left* to *right*. (Reproduced from ref. 18 with permission from American Society of Plant Biologists.)

could result in lethality (18). Thus, knockdown lines are appropriate for investigating the function of a targeted gene. *Ostrxm* RNAi plants exhibited severe developmental defects, including semi-dwarfism and pale green color (18). An example of the results produced is shown in **Fig. 15.1**. Since thioredoxin *m* has similar members in its gene family, it was important to investigate cross-interference of other family members by RT-PCR and Northern blot using specific primers and probes, respectively, in order to verify correct targeting. The critical cutoff point for cross-interference in RNAi has been proposed to be between 87 and 67% (21). The other two members of the thioredoxin *m* family found in rice have 23.9 and 24.2% nucleotide sequence homology, respectively, with our target; however, we did not detect cross-interference by RT-PCR in the surviving lines (18). Through 2-DE and MALDI-TOF MS analysis, high-throughput protein profiling of genetically modified organisms can be carried out. These techniques enable investigation of the in vivo function and molecular interactions of the protein target together with the phenotype of the organism. A description of the methods and procedures developed and carried out for this type of analysis follows.

3.1. Preparation of the Ostrxm hpRNA Vector

1. The first intron of *OsCDPK* (1 kb, accession number AF 194414) was amplified by PCR using rice genomic DNA as a template and then ligated into the *Pst*I/*Eco*RI restriction sites of pBluescript KS (–) (Stratagene).
2. An open reading frame (ORF) (519 base pairs) of the *Ostrxm* cDNA containing *Bam*HI/*Pst*I restriction sites at either end was inserted upstream of the *OsCDPK* intron in the antisense orientation.
3. An ORF of the *Ostrxm* cDNA containing *Eco*RI/*Cla*I restriction sites at either end was inserted downstream of the intron, in the sense orientation. The complete *Ostrxm* hpRNA construct contained an antisense fragment of the *Ostrxm* cDNA, followed by the *OsCDPK* intron and then a sense fragment of the *Ostrxm* cDNA, including a stop codon.
4. The *Ostrxm* hpRNA construct was inserted into the *Bam*HI/*Cla*I sites of pCAMBIA1300, between the CaMV35S promoter and the 3′-flanking region of the octopine synthase gene (*Ocs* terminator) to generate pCAMBIA1300::*Ostrxm*. An example of the results produced is shown in **Fig. 15.1a**.
5. The *Ostrxm* hpRNA plasmid was then transformed into rice using *Agrobacterium* (*Agrobacterium tumefaciens* strain LBA4404)-mediated transformation (*see* **Sections 3.2**, **3.3**, and **3.4**).

3.2. Sterilization of Rice Seeds

1. Rice seeds were carefully hulled by hand and put into a 50 mL conical tube.
2. Seeds were rinsed using two to three seed volumes of 70% ethanol for 5 min each with shaking at 130 rpm.
3. Ethanol was removed by washing two to three times with sterile dH_2O.
4. Seeds were sterilized by incubation in 3–6% sodium hypochlorite for 15 min, with shaking at 130 rpm. The procedure was repeated twice, and then residual sodium hypochlorite was removed by washing 10 times with sterile dH_2O.
5. Residual water was removed by drying the seeds on sterile filter paper in a petri dish.
6. Sixteen sterilized seeds were placed on a plate containing MSJ4 medium; embryos were placed in an upward (45–60°) orientation.
7. Plates were sealed with a stretchable film and incubated at 28°C in the dark for 3 weeks.

3.3. Induction of Embryogenic Calli and Agrobacterium-Mediated Rice Transformation

1. Root and shoot radicles were excised from seeds after the 3-week growth period.
2. Remaining tissues were transferred to new MSJ4 medium and incubated at 28°C for 2 weeks.
3. Round glossy embryogenic calli were selected and transferred to MSJ4 medium (*see* **Note 5**). The plate was incubated at 28°C in the dark for 3–4 days.
4. Prior to plant transformation, Agrobacteria carrying the pCAMBIA1300 binary vector containing the *Ostrxm* hpRNA cassette in the T-DNA region were inoculated into 5 mL of YEP liquid culture medium and then cultured in a 28°C incubator with shaking.
5. After 2 days, 1 mL of the precultured *Agrobacterium* was inoculated into 250 mL of YEP medium and then incubated at 28°C until the optical density at 600 nm (OD_{600}) reached between 1.0 and 1.2.
6. Agrobacteria were harvested by centrifugation at room temperature for 5 min at 5,000×*g*.
7. The pellet was resuspended in an equal volume of inoculation medium (AA medium).
8. The precultured embryogenic calli were transferred to an empty petri dish and treated with the *Agrobacterium* suspension for 1 h at 28°C in the dark.
9. Calli were removed from the *Agrobacterium* suspension and dried on sterile filter paper.
10. Calli were transferred to solid MSJ2 medium close together and incubated at 25°C for 3 days in the dark.

3.4. Selection of Transformed Calli and Regeneration of the Transformed Callus

1. Agrobacteria from the cocultivated calli were transferred into a 50 mL conical tube and then washed three to five times using sterile dH_2O containing cefotaxime (250 mg/L).
2. Sets of 25 calli were transferred to plates containing MSJ4 medium and then incubated at 28°C for 2–3 weeks in the dark.
3. Calli were transferred to fresh MSJ4 medium and then incubated at 28°C for 2–3 weeks in the dark.
4. Calli that were growing vigorously on the selection medium were selected and transferred to MSR medium.
5. Transferred calli were incubated at 28°C in the light and subcultured on new MSR medium every 2–3 weeks until regenerated shoots appeared.
6. Shoots grown to 5 cm were acclimated in a 1% hyponex solution for 7–10 days and then transferred to soil.

3.5. Total RNA Preparation

1. One gram of rice leaf tissue was homogenized in 1.8 mL of RNA extraction buffer using a mortar and pestle, and then the homogenate was transferred to two Eppendorf tubes.
2. Samples were mixed by vortexing and then the pellet was precipitated at 4°C for 5 min at 15,000×*g*.
3. The supernatant (1.2 mL) was carefully applied to the top of a 750 μL, 5.7 M, CsCl cushion in an ultracentrifuge tube.
4. Samples were subjected to centrifugation at 20°C for 14 h at 100,000×*g* using a Beckman TLA45 ultracentrifuge.
5. The buffer phase and cushion were removed (*see* **Note 6**) and washed with DEPC-treated H_2O, and then the pellet was dissolved in 200 μL of TE buffer containing 1% SDS.
6. The dissolved RNA was extracted once with chloroform/1-butanol (4:1) and then precipitated overnight following the addition of 1/10 volume of 3 M sodium acetate (NaOAc), pH 5.2, and 2.5 volumes of ethanol.
7. RNA was recovered by centrifugation at 4°C for 15 h at 15,000×*g*.
8. The pellet was dissolved in 200 μL of TE buffer and the concentration of RNA was determined by measuring absorbance at 260 nm.

3.6. Northern Blot Analysis

1. Fifteen micrograms of total RNA was separated by 1.2% formaldehyde agarose gel electrophoresis; ribosomal (r) RNA bands were used as loading controls.
2. RNA was transferred to a nylon membrane overnight and then cross-linked by exposure to UV light for 2 min; 10X SSC was used as the transfer buffer.
3. The blot was hybridized by incubation with [α-^{32}P]ATP-labeled *Ostrxm* cDNA as the probe. An example of the results produced is shown in **Fig. 15.2a**.

3.7. RT-PCR Analysis of RNAi

1. One microgram total RNA template and 500 ng oligo $(dT)_{18}$ primer were added into a sterile, nuclease-free tube on ice and volume of the mixture was adjusted to 12 μL by adding DEPC-treated water (*see* **Note 7**).
2. Four microliter 5X reaction buffer, 1 μL RNase inhibitor (20 U/μL; Fermentas), 2 μL 10 mM dNTP mix, and 2 μL M-MuLV reverse transcriptase (20 U/μL; Fermentas) were added into the template and primer mixture. The final reaction volume was 20 μL.
3. First-strand cDNA reaction was performed at 42°C for 50 min (*see* **Note 8**).
4. The reaction was terminated by inactivating the enzyme at 70°C for 15 min.

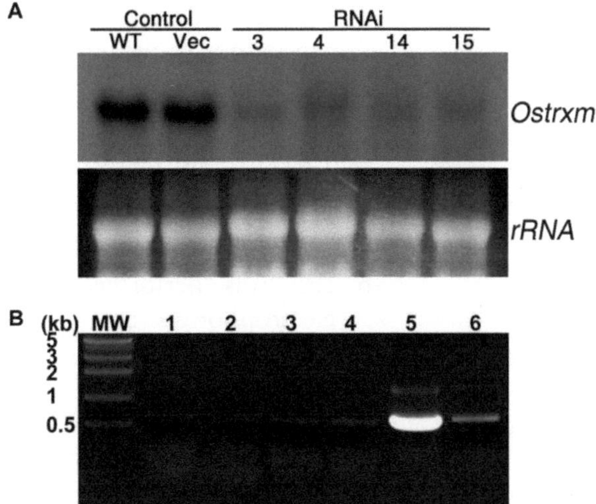

Fig. 15.2. Analysis of *Ostrxm* expression in transgenic rice plants and expression of rice thioredoxin *m* isoforms. (**a**) Northern blot analysis of the T_1 progeny of wild type (WT), pCAMBIA 1300 vector transformed (Vec), and *Ostrxm* RNAi transgenic plants using [α-^{32}P]ATP-labeled *Ostrxm* cDNA as a probe. (**b**) Quantification of genes encoding rice thioredoxin *m* isoforms. Total RNA was prepared from leaves of 12-day-old WT (lanes 1, 3, 5) and *Ostrxm* RNAi #14 (*lanes 2, 4, 6*) rice seedlings. RT-PCR was performed using specific oligonucleotide primers for Os02g42700 (*lanes 1, 2*), Os04g44830 (*lanes 3, 4*), and Os12g08730 (*lanes 5, 6*), as described in **Sections 2** and **3**. (Reproduced from ref. 18 with permission from American Society of Plant Biologists.)

5. Two microliters of the first-strand cDNA was used as a template for the PCR in a total volume of 50 μL containing 200 nM each PCR primer, 200 μM each dNTP, and 2.5 units of *Taq* DNA polymerase in 20 mM Tris–HCl (pH 8.4), 50 mM KCl, and 1.5 mM $MgCl_2$.
6. PCR was carried out using a PTC-0220 PCR machine (MJ Research). The thermal cycling parameters were as follows: denaturation at 95°C for 5 min, followed by 25–35 cycles of 95°C for 1 min, 60°C for 30 s, and 72°C for 1 min.
7. Detection of the low-level expression genes Os02g42700 and Os04g44830 required more than 30 cycles, whereas 25 cycles were sufficient to detect Os12g08730 (the *Ostrxm* target of the study).
8. Ten microliters of each PCR product was analyzed on a 1.2% TAE-agarose gel stained with 5 μg/mL ethidium bromide. An example of the results produced is shown in **Fig. 15.2b**.

3.8. Protein Sample Preparation, Isoelectric Focusing (IEF), and 2-DE

1. Two grams of rice leaves ground in liquid nitrogen were homogenized in 10 mL of sample lysis buffer.
2. Proteins were extracted for 1 h at room temperature with vortexing (*see* **Note 9**).

3. Protein loading was normalized according to protein concentration, as measured by the Bradford assay (22).
4. IPG dry strips were equilibrated for 12–16 h in a solution of 7 M urea, 2 M thiourea containing 2% CHAPS, 1% DTT, and 1% pharmalyte.
5. Samples (200 μg each) were loaded on the equilibrated strips for separation in the first dimension.
6. IEF was performed at 20°C using a Multiphor II electrophoresis unit and EPS 3500 XL power supply (Amersham Biosciences), according to the manufacturer's instructions (*see* **Note 10**).
7. Strips were incubated for 10 min each in equilibration buffer with 1% DTT followed by equilibration buffer with 2.5% iodoacetamide.
8. Equilibrated strips were inserted into SDS-PAGE gels (20 × 24 cm, 10–16%).
9. SDS-PAGE was performed using a Hoefer DALT 2-D system (Amersham Biosciences) according to the manufacturer's instructions (*see* **Note 11**).
10. Proteins were visualized using silver staining, omitting the fixing and sensitization step with glutaraldehyde (23) (*see* **Note 12**). An example of the results produced is shown in **Fig. 15.3**.
11. Quantitative analysis of digitized images was carried out using PDQuest (version 7.0, BioRad) software, according to the manufacturer's instructions (*see* **Note 13**).
12. Protein spots were selected for further analysis if expression deviated over twofold as compared to the control.

3.9. In-Gel Protein Digestion

1. In-gel proteolysis of protein spots was carried out using modified porcine trypsin (24).
2. Gel pieces were washed with 50% acetonitrile to remove SDS, salt, and stain; dried to remove solvent; and then rehydrated in 50 μL trypsin (8–10 ng/μL in 40 mM ammonium bicarbonate with 10% acetonitrile, pH 8.1) and incubated for 8–10 h at 37°C.
3. Digestion was terminated by the addition of 5 μL of 0.5% trifluoroacetic acid.
4. Tryptic peptides were recovered by combining the aqueous phases from several rounds of extraction of the gel pieces by agitation with 50% aqueous acetonitrile.
5. The peptide mixture was concentrated and then desalted using C_{18}ZipTips (Millipore); peptides were eluted in 1–5 μL of acetonitrile.

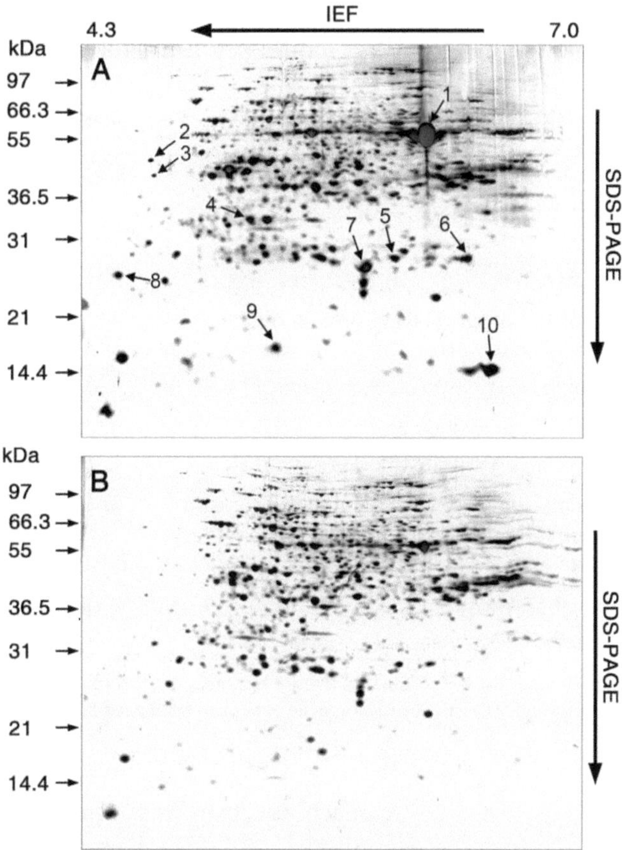

Fig. 15.3. 2-DE analysis of differentially expressed proteins in *Ostrxm* RNAi plants. Protein samples were separated by 2-DE (p*I* 4–7) and then visualized by silver staining. Proteins whose expression was decreased in *Ostrxm* RNAi plants are indicated by *arrows*. WT (**a**) and *Ostrxm* RNAi (**b**) plants. Relative molecular weights are indicated on the *left side* of each panel in kDa. The number of each protein spot (1–10) corresponds to the listing in **Table 15.1**. Thirty-day-old rice plants grown in a greenhouse (sunlight) were used for the analysis. (Reproduced from ref. 18 with permission from American Society of Plant Biologists.)

6. An aliquot of the eluate was mixed with an equal volume of a saturated solution of α-cyano-4-hydroxycinnamic acid in 50% aqueous acetonitrile, and then 1 μL of the mixture was spotted onto a target plate.

3.10. MALDI-TOF MS Analysis and Database Search

1. MALDI-TOF MS was performed using an Ettan MALDI-TOF system (Amersham Biosciences).
2. Peptides were evaporated with a nitrogen (N_2) laser at 337 nm using a delayed extraction approach.
3. Samples were accelerated with a 20 Kv injection pulse for the time-of-flight analysis.

Table 15.1
Identification of differentially expressed proteins in *Ostrxm* RNAi plants

Spot	Protein	Accession no.	Est'd Z	pI/MW
1	Ribulose bisphosphate carboxylase large chain	CAG34174	1.15	6.2/53.35
2	ND			
3	Peptidyl-prolyl *cis–trans* isomerase	BAD05657	2.31	4.8/46.81
4	33 kDa oxygen evolving protein of photosystem II	BAB64069	2.39	5.2/30.10
5	Ribulose-5-phosphate-3-epimerase	AAL84303	0.99	5.7/25.02
6	Carbonic anhydrase 3	AAD56038	2.24	6.3/22.30
7	23 kDa polypeptide of photosystem II	AAC98778	2.13	6.0/23.33
8	Putative benzothiadiazole-induced somatic embryogenesis receptor kinase 1	BAD53863	1.15	5.6/44.21
9	ND			
10	Ribulose-1,5-bisphosphate carboxylase/oxygenase small subunit	AAR19268	1.91	6.8/15.05

ND, not determined; Est'd Z, estimated *Z* score as a measure of confidence for Profound-based identification. Reproduced from ref. 18 with permission from American Society of Plant Biologists

4. Each spectrum was the cumulative average of 300 laser shots.
5. The search program ProFound, developed at Rockefeller University (http://129.85.19.192/profound_bin/WebProFound.exe), was used for protein identification by peptide mass fingerprinting.
6. Spectra were calibrated with trypsin auto-digestion ion peaks *m/z* (842.510, 2211.1046) as internal standards. An example of the results produced is shown in **Table 15.1**.

4. Notes

1. All culture media should be freshly prepared by diluting the individual stock solutions using deionized water before use and sterilized by autoclaving at 121°C for 10–15 min.
2. Stock solution should be stored in tightly sealed glassware to avoid alteration of the initial concentration due to evaporation and can be used for a month.
3. Acetosyringone should be added at a final concentration of 200 μM after sterilization of the medium.

4. β-mercaptoethanol and sarkosyl should be added just before the extraction.
5. Approximately 300–400 calli were transferred on a plate close together.
6. Carefully pour off the supernatant from the tube and then completely remove the remaining liquid with a pipet.
7. If the RNA template is GC-rich or contains secondary structures, the template and oligo $(dT)_{18}$ primer mixture was incubated at 65°C for 5 min and chilled on ice, optionally.
8. The reaction temperature can be increased up to 45°C for GC-rich RNA templates.
9. After centrifugation at 15,000×*g* for 1 h at 15°C, the insoluble material was discarded and the soluble fraction was collected for 2-DE analysis.
10. Voltage was linearly increased from 150 to 3,500 V over 3 h to allow sample entry, followed by a constant voltage of 3,500 for 26 h for a total of 96 kVh.
11. Separation in the second dimension was carried out at 20°C for 1,700 Vh.
12. For subsequent analysis by MALDI-TOF MS, the fixing and sensitization step using glutaraldehyde should be omitted from the silver staining protocols, since this cross-linking reagent chemically modifies the protein and inhibits subsequent steps.
13. The intensity of each spot was normalized to total valid spot intensity.

Acknowledgments

This work was supported by grants from MEST/NRF to the WCU program (R32-10148), CFGC (CG3313-1) and NRL (M10600000205-06J0000-20510), and Technology Development Program for Agriculture and Forestry (609004-5) of ARPC, MIFAFF, Korea.

References

1. Holmgren, A. (1989) Thioredoxin and glutaredoxin systems. *J. Biol. Chem.* **264**, 13963–13966.
2. Balmer, Y., and Buchanan, B. B. (2002) Yet another plant thioredoxin. *Trends Plant Sci.* 7, 191–193.
3. Collin, V., Lamkemeyer, P., Miginiac-Maslow, M., Hirasawa, M., Knaff, D. B., Dietz, K. J., and Issakidis-Bourguet, E. (2004) Characterization of plastidial thioredoxins from *Arabidopsis* belonging to the new y-type. *Plant Physiol.* **136**, 4088–4095.

4. Gelhaye, E., Rouhier, N., and Jacquot, J. P. (2004) The thioredoxin h system of higher plants. *Plant Physiol. Biochem.* **42**, 265–271.
5. Collin, V., Issakidis-Bourguet, E., Marchand, C., Hirasawa, M., Lancelin, J. M., Knaff, D. B., and Miginiac-Maslow, M. (2003) The *Arabidopsis* plastidial thioredoxins: New functions and new insights into specificity. *J. Biol. Chem.* **278**, 23747–23752.
6. Mestres-Ortega, D., and Meyer, Y. (1999) The *Arabidopsis thaliana* genome encodes at least four thioredoxins m and a new prokaryotic-like thioredoxin. *Gene* **240**, 307–316.
7. Ruelland, E., and Miginiac-Maslow, M. (1999) Regulation of chloroplast enzyme activities by thioredoxins: Activation or relief from inhibition? *Trends Plant Sci.* **4**, 136–141.
8. Clancey, C. J., and Gilbert, H. F. (1987) Thiol/disulfide exchange in the thioredoxin-catalyzed reductive activation of spinach chloroplast fructose-1,6-bisphosphatase. Kinetics and thermodynamics. *J. Biol. Chem.* **262**, 13545–13549.
9. Droux, M., Jacquot, J. P., Miginac-Maslow, M., Gadal, P., Huet, J. C., Crawford, N. A., Yee, B. C., and Buchanan, B. B. (1987) Ferredoxin-thioredoxin reductase, an iron-sulfur enzyme linking light to enzyme regulation in oxygenic photosynthesis: Purification and properties of the enzyme from C3, C4, and cyanobacterial species. *Arch. Biochem. Biophys.* **252**, 426–439.
10. Droux, M., Miginiac-Maslow, M., Jacquot, J. P., Gadal, P., Crawford, N. A., Kosower, N. S., and Buchanan, B. B. (1987) Ferredoxin-thioredoxin reductase: A catalytically active dithiol group links photoreduced ferredoxin to thioredoxin functional in photosynthetic enzyme regulation. *Arch. Biochem. Biophys.* **256**, 372–380.
11. Scheibe, R., and Anderson, L. E. (1981) Dark modulation of NADP-dependent malate dehydrogenase and glucose-6-phosphate dehydrogenase in the chloroplast. *Biochim. Biophys. Acta.* **636**, 58–64.
12. Geck, M. K., Larimer, F. W., and Hartman, F. C. (1996) Identification of residues of spinach thioredoxin f that influence interactions with target enzymes. *J. Biol. Chem.* **271**, 24736–24740.
13. Hodges, M., Miginiac-Maslow, M., Decottignies, P., Jacquot, J. P., Stein, M., Lepiniec, L., Cretin, C., and Gadal, P. (1994) Purification and characterization of pea thioredoxin f expressed in *Escherichia coli. Plant Mol. Biol.* **26**, 225–234.
14. Muller, E. G., and Buchanan, B. B. (1989) Thioredoxin is essential for photosynthetic growth. The thioredoxin *m* gene of *Anacystis nidulans. J. Biol. Chem.* **264**, 4008–4014.
15. Duek, P. D., and Wolosiuk, R. A. (2001) Rapeseed chloroplast thioredoxin-m. Modulation of the affinity for target proteins. *Biochim. Biophys. Acta.* **1546**, 299–311.
16. Issakidis-Bourguet, E., Mouaheb, N., Meyer, Y., and Miginiac-Maslow, M. (2001) Heterologous complementation of yeast reveals a new putative function for chloroplast m-type thioredoxin. *Plant J.* **25**, 127–135.
17. Macrae, I., Zhou, K., Li, F., Repic, A., Brooks, A., Cande, W., Adams, P., and Doudna, J. (2006) Structural basis for double-stranded RNA processing by dicer. *Science* **311**, 195–198.
18. Chi, Y. H., Moon, J. C., Park, J. H., Kim, H. S., Zulfugarov, I. S., Fanata, W. I., Jang, H. H., Lee, J. R., Lee, Y. M., Kim, S. T., Chung, Y. Y., Lim, C. O., Kim, J. Y., Yun, D. J., Lee, C. H., Lee, K. O., and Lee, S. Y. (2008) Abnormal chloroplast development and growth inhibition in rice thioredoxin *m* knock-down plants. *Plant Physiol.* **148**, 808–817.
19. Murashige, T., and Skoog, F. (1962) A revised medium for rapid growth and bioassays with tobacco tissue cultures. *Physiol. Plant.* **15**, 473–497.
20. Nitsch, J. P., and Nitsch, C. (1969) Haploid plants from pollen grains. *Science* **163**, 85–87.
21. Ui-Tei, K., Zenno, S., Miyata, Y., and Saigo, K. (2000) Sensitive assay of RNA interference in *Drosophila* and Chinese hamster cultured cells using firefly luciferase gene as target. *FEBS Lett.* **479**, 79–82.
22. Bradford, M. (1976) A rapid and sensitive method for the quantitation of microgram quantities of protein utilizing the principle of protein-dye binding. *Anal. Biochem.* **72**, 248–254.
23. Okaley, B. R., Kirsch, D. R., and Morris, R. (1980) A simplified ultrasensitive silver stain for detecting proteins in polyacrylamide gels. *Anal. Biochem.* **105**, 361–363.
24. Shevchenko, A., Wilm, M., Vorm, O., and Mann, M. (1996) Mass spectrometric sequencing of proteins from silver stained polyacrylamide gels. *Anal. Chem.* **68**, 850–858.

Chapter 16

Comparative Analysis of Phosphoprotein Expression Using 2D-DIGE

Tomoya Asano and Takumi Nishiuchi

Abstract

Two-dimensional gel electrophoresis (2-DE) has often been used to compare protein expression pattern between different samples. This method is also useful to compare protein expression between wild-type and RNAi plants. 2D-DIGE (difference gel electrophoresis) was developed to perform quantitative proteomics of two or more samples. In this method, proteins are fluorescently labeled by Cy2, Cy3, or Cy5 and are mixed and subjected to electrophoresis on the same gel, thereby gel-to-gel variations are eliminated. We perform phosphoproteomics between *Arabidopsis* wild-type and a stress-activated MAPK (mitogen-activated protein kinase) mutant after the phytotoxin treatment. For this purpose, proteins are extracted from both samples, and then phosphorylated proteins are purified by phosphoprotein enrichment columns. Purified proteins are fluorescently labeled by different CyDyes using the minimum labeling method. The labeled protein samples are separated on the same gel by 2-DE, and gels are scanned by a variable mode laser imager. Then, high-resolution images are analyzed by 2D analysis software. Thus, we identified several phosphoprotein spots that were differentially accumulated between wild-type and *mapk* mutant.

Key words: 2D-DIGE, MAPK, phosphoprotein, proteome.

1. Introduction

Two-dimensional gel electrophoresis (2-DE) has a long history and is frequently used to profile protein expression pattern in many organisms. This method is also useful to compare protein expression between wild-type and RNAi plants. Protein preparation is the first and important step to the success of 2-DE. Methods of protein preparation vary depending on the purpose such as posttranslationally modified proteins and organelle proteins (1, 2). The composition of protein extraction buffer is known

H. Kodama, A. Komamine (eds.), *RNAi and Plant Gene Function Analysis*, Methods in Molecular Biology 744,
DOI 10.1007/978-1-61779-123-9_16,

to be reflected in the patterns of 2-DE. Therefore, we precipitate protein extractions with trichloroacetic acid (TCA), then extraction buffers were replaced by buffer for 2-DE. Nevertheless, some substances such as ionic detergents might be reflected in the patterns of 2-DE. The composition of the protein extraction buffer needs to be explored.

In the conventional 2-DE, electrophoresed gels are generally subjected to colloidal CBB (Coomassie brilliant blue) staining or fluorescent staining such as deep purple (GE Healthcare). Since it is necessary to compare protein spots between different gel images, gel-to-gel variations often bother 2-DE users. 2D difference gel electrophoresis (DIGE) clears this problem, since multiple protein samples labeled by different fluorescent dyes are separated by the same gel electrophoresis (3, 4). Therefore, 2D-DIGE has higher reproducibility and reliability than the conventional 2-DE. In minimum labeling method (GE Healthcare), *N*-hydroxysuccinimidyl (NHS) esters of the CyDyes are incorporated into the ϵ-amino acid group of lysine (4). Only one lysine of each protein should be labeled for reproducibility and reliability of DIGE method (5). In this step, pH of the reaction solution has to be maintained at 8–9. One sample is labeled with Cy3 and the other with Cy5. A third dye, Cy2, is used to label a mixture of all the samples, which becomes an internal standard. Alternatively, additional samples can be labeled with Cy2. Equal amounts of these three labeled protein mixtures are combined and are run on the same 2-DE, eliminating any gel-to-gel variation.

After electrophoresis, gels with glass plates are scanned with variable mode laser imager such as Typhoon (GE Healthcare). Then, resulting images are analyzed by 2D analysis software such as DeCyder (GE Healthcare) and PDQuest (Bio-Rad). A spot picker needs to excise protein spots from gels of 2D-DIGE. CyDye labeling causes the increase of molecular mass of original protein (6). Low efficiency of CyDye protein labeling makes it difficult to identify protein spots by mass spectrometry analysis. Alternatively, same samples for 2D-DIGE are subjected to conventional 2-DE, then visible protein spots are picked up for mass spectrometry analysis. In this chapter, we perform phosphoproteomics to identify phosphorylated proteins mediated by the stress-activated mitogen-activated protein kinases (MAPKs). Some phytopathogens such as *Fusarium* species produce trichothecene phytotoxins (7). T-2 toxin is one of the trichothecene phytotoxin and has an elicitor-like activity in *Arabidopsis* plants (8). In fact, T-2 toxin activates some stress-activated MAPKs in *Arabidopsis*. To investigate proteins phosphorylated by the MAPK, we purify phosphorylated proteins from wild-type and the *mapk* mutant after T-2 toxin treatment by phosphoprotein enrichment columns. Then, proteome analysis is carried out

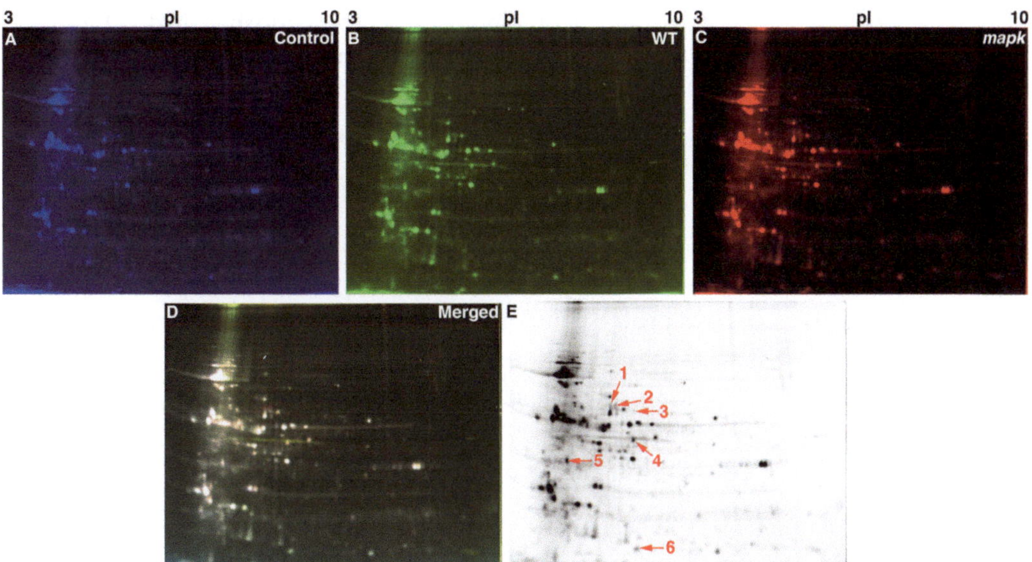

Fig. 16.1. 2D-DIGE analysis between WT and the *mapk* mutant by T-2 toxin treatment. Phosphoproteins from internal control (**A**), WT (**B**), and *mapk* mutant samples (**C**) were labeled by Cy2, Cy3, and Cy5, respectively. Merged images of (**A**), (**B**), and (**C**) are shown in (**D**). Selected spots (**E**) are differentially accumulated between WT and the *mapk* mutant.

Table 16.1
Several phosphoprotein spots were differentially accumulated between wild-type and the *mapk* mutant

Spot no.	WT	*mapk*	WT/*mapk*
1	3,824.2	38,516.9	0.099286287
2	67,561.1	207,749.8	0.325204164
3	4,042.2	19,371.1	0.208671681
4	109,272.5	145,280.6	0.752147912
5	110,949.4	146,577.1	0.756935429
6	148,441.9	110,178.2	1.34728921

by 2D-DIGE (**Fig. 16.1**). In the experiment, we revealed that amounts of several proteins were downregulated or upregulated in the *mapk* mutant compared with wild type (**Table 16.1**).

2. Materials

2.1. Plant Growth and Trichothecene Phytotoxin Treatment

1. Murashige and Skoog (MS) agar medium: Murashige and Skoog plant salt mixture (Wako, Osaka, Japan), 0.1% vitamin stock mixture (3 g/L thiamine hydrochloride, 5 g/L

nicotinic acid, and 0.5 g/L pyridoxine hydrochloride), 2% sucrose, 0.2% gellan gum, adjusted to pH 5.8 by 0.1 N KOH.

2. T-2 toxin stock solution (10 mg/mL) is dissolved in 99.5% ethanol (*see* **Note 1**).

2.2. Phosphoprotein Preparation

1. Protein extraction buffer: HEPES buffer containing 1% Triton X-100 (*see* **Note 2**), 1 mM PMSF, and 1/1,000 volume of protease inhibitor cocktail (Sigma-Aldrich).
2. Pro-Q Diamond Phosphoprotein Enrichment Kit (Molecular Probes) including columns, wash buffer, and elution buffer.
3. Vivaspin 500 centrifugal concentrator (GE Healthcare).
4. 100% TCA. Store at 4°C.
5. 100% acetone. Store at room temperature.
6. RC DC Protein Assay Kit (Bio-Rad).

2.3. Protein Labeling

1. Lysis buffer: 7 M urea, 2 M thiourea, 4% 3-[(3-cholamidopropyl) dimethylammonio]-1-propanesulfonate (CHAPS) and 30 mM Tris–HCl, pH 8.5.
2. CyDye DIGE Fluors (GE Healthcare).
3. Rehydration buffer: 7 M urea, 2 M thiourea, 4% CHAPS, 0.005% bromophenol blue, 10 mM DTT, and 0.5% IPG buffer (GE Healthcare).
4. 10 mM lysine.

2.4. 2D Electrophoresis

1. 30% acrylamide/*N,N′*-methylenebisacrylamide solution (29.2:0.8). Store at 4°C in the dark.
2. *N,N,N,N′*-tetramethylethylenediamine (TEMED). Store at 4°C.
3. 1 M Tris–HCl, pH 8.8. Store at room temperature.
4. 10% SDS solution. Store at room temperature.
5. 25% ammonium persulfate is dissolved in water and is stored at –20°C.
6. 10X running buffer: 2.5 M Tris base, 19.2 M glycine, and 10% SDS.
7. Equilibration buffer 1: 50 mM Tris–HCl, pH 8.8, 6 M urea, 30% (v/v) glycerol, 2% (w/v) SDS, 1% DTT, 0.002% (w/v) bromophenol blue.
8. Equilibration buffer 2: 50 mM Tris–HCl, pH 8.8, 6 M urea, 30% (v/v) glycerol, 2% (w/v) SDS, 2.5% iodoacetamide, 0.002% (w/v) bromophenol blue.

2.5. In-Gel Digestion for MALDI-TOF/TOF Analysis

1. 1 M ammonium bicarbonate stock solution.
2. Iodoacetamide. Store at room temperature.
3. Sequencing-grade trypsin (Promega). Store at –20°C.
4. 90% formic acid solution.
5. ZipTipC18 columns (Millipore).
6. Matrix solution (store at –20°C): 20 mg α-cyano-4-hydroxycinnamic acid (αCHCA), 0.4 mg diammonium citrate, 5 μL trifluoroacetic acid (TFA), and 3.5 mL CH_3CN per 5 mL.

3. Methods

3.1. Preparation of Samples

1. The Columbia (Col-0) ecotype of *Arabidopsis thaliana* (L.) Heynh is used as a wild type in this study. A T-DNA insertion mutant of *mapk* mutant is obtained from the *Arabidopsis* Biological Resource Center, Ohio state University, Columbus, OH.
2. Surface-sterilized seeds are sown on MS agar medium in disposable plastic petri dishes.
3. The seeds are placed at 4°C under dark condition for 2 days and then grown at 22°C under a 16/8-h light/dark cycle.
4. For T-2 toxin treatments, 10-day-old *Arabidopsis* plants grown on MS medium without trichothecene are transferred to 0.1 μM T-2 toxin-containing medium.

3.2. Protein Extraction and Purification of Phosphorylated Protein

1. *Arabidopsis* shoots are ground to a fine powder with pestles in liquid nitrogen. For protein extraction, about 2 g of these fine powders is mixed with 5 mL of protein extraction buffer and then are homogenized with a vortex mixer.
2. These extraction solutions are centrifuged (14,000×*g*, 4°C, 15 min), and the supernatant is collected. Protein contents in these solutions are measured by RC DC Protein Assay Kit (Bio-Rad, USA) (*see* **Note 3**).

3.3. Purification of Phosphorylated Protein and Protein–Cyanine Dye Labeling

1. Dilute the protein extractions to a final protein concentration of 0.1–0.2 mg/mL with wash buffer (*see* **Section 2.2**, Step 2). Equilibrate the Pro-Q Diamond Phosphoprotein Enrichment Columns (Molecular Probes) with 1 mL of wash buffer twice.

2. Apply the appropriate volume (maximum input 1 mg) of diluted protein solutions to the column and then wash the column with 1 mL of wash buffer three times.
3. Elute the phosphoproteins from the columns with 250 μL of elution buffer five times, collect the elute fractions, and concentrate up to approximately 50 μL by the vivaspin 500 concentrator. When 1 mg protein extracted from *Arabidopsis* leaves is loaded onto the column, approximately 50 μg phosphoprotein is recovered.
4. Purified phosphoproteins are precipitated by 5% TCA (final concentration) and washed by 100% acetone at least twice. The phosphorylated proteins are dried by aspirator for 5 min.
5. The dried precipitates are dissolved in lysis buffer. Insoluble material is removed by centrifugation (15,000×*g*, 4°C, 15 min). Check the pH (from 8 to 9) of the supernatants by litmus papers.
6. Protein labeling was performed using the CyDye DIGE Fluors (GE Healthcare; *see* **Note 4**). Proteins from the control sample (Sample 1: 1/2 Sample 2 and 1/2 Sample 3), T-2 toxin-treated samples of wild type (Sample 2), and T-2 toxin-treated samples of the *mapk* mutant (Sample 3) were covalently labeled with Cy2, Cy3, and Cy5, respectively.
7. Thirty microgram proteins can be labeled with 200 pmol of amine-reactive CyDyes. The labeling mixture was incubated on ice in the dark for 30 min. To stop labeling reaction, add 1 μL of 10 mM lysine, which reacts with the free NHS esters of the CyDyes and incubate on ice for 10 min. Each of the labeled protein samples was mixed and rehydration buffer is added to samples up to 340 μL.

3.4. 2D Electrophoresis

1. For isoelectric focusing (IEF), CyDye-labeled samples were loaded onto rehydration strips of 18 cm in length with immobilized pH gradient from 3 to 10 and separated on a MultiPhor Unit (GE Healthcare). Increase the voltage linearly from 300 to 3,500 V during 1.5 h, followed by additional 5 h at 3,500 V.
2. After IEF, the strips are equilibrated for 15 min with equilibration buffer 1.
3. They are equilibrated for 15 min with equilibration buffer 2.
4. Prepare a 10% SDS-polyacrylamide gel. Mix 10 mL of 30% acrylamide/*N*,*N*′-methylenebisacrylamide solution, 8.4 mL of deionized water, 11.2 mL of 1 M Tris–HCl, pH 8.8, 300 μL of 10% SDS solution, 24 μL of TEMED, and finally add 100 μL of 25% ammonium persulfate. Then, the mixture is poured into gel plates.

5. The strips (from Step 3) are mounted on the top of a 10% SDS-polyacrylamide gel with stacking gel in a Hoefer SE 600 Ruby system (GE Healthcare). Electrophoresis is carried out at 10 mA constant current for 18 h.

3.5. Gel Imaging and Data Analysis

1. After electrophoresis, fluorescent signals are detected by a variable Typhoon™ 9400 imager (GE Healthcare).
2. Cy2 signals are scanned using a 488 nm laser and an emission filter of 520 nm band pass (BP) 40. Cy3 signals are scanned using a 532 nm laser and an emission filter of 580 nm BP 30. Cy5 signals are scanned using a 633 nm laser and an emission filter of 670 nm BP 30. All gels are scanned at 100 μm (pixel size) resolution. The photomultiplier tube (PMT) is set at 600 V with normal sensitivity.
3. The scanned images are directly transferred to the ImageQuant V5.2 software package (GE Healthcare).
4. Gel images are saved as 16-bit TIF files by Adobe Photoshop CS3 software. Unnecessary regions of gel images are removed by Adobe Photoshop CS3. For quantification of protein spots, these images are saved as grayscale files. Resulting images are shown in **Fig. 16.1**.
5. Comparison of protein spots among these images is performed by PDQuest Advanced (Bio-Rad; **Table 16.1**).
6. To identify protein spots, spots are excised from gels by Ettan spot picker (GE Healthcare).

3.6. Protein Identification by MALDI-TOF/TOF Analyzer

1. Protein spots (gels) are cut into small pieces and then are dehydrated by 100 μL of 100% CH_3CN for 30 min in 1.5 mL tubes.
2. After centrifuge, supernatants are removed. The samples are reduced by 100 μL of 10 mM DTT in 100 mM ammonium bicarbonate for 1 h at 56°C and then are alkylated with 100 μL of 32.4 mM iodoacetamide in 100 mM ammonium bicarbonate buffer for 45 min in the dark (*see* **Note 5**).
3. After centrifuge, supernatants are removed. Gel pieces are washed alternately by 100 μL of 100 mM ammonium bicarbonate buffer and 100 μL of 100% CH_3CN. Repeat this step twice. Then, the samples are digested with 20 μL of 12.5 ng/μL sequencing-grade trypsin (Promega) in 20 mM ammonium bicarbonate buffer overnight at 37°C. If the buffer is insufficient, add appropriate volume of 20 mM ammonium bicarbonate buffer to the tubes.
4. Add 20 μL of 20 mM ammonium bicarbonate buffer to the digested peptides and then mix well for 10 min.

5. Spin down briefly. Then, supernatants are transferred to collection tubes including 100 μL of 50% CH_3CN and 5% formic acid solution.
6. The digested peptides are eluted from gels by 100 μL of 50% CH_3CN and 5% formic acid solution for 30 min. Spin down briefly. The supernatant is transferred to the collection tubes (*see* **Section 3.6**, Step 5). Repeat this step twice. Using a centrifugal concentrator, the elution (total ~340 μL) is concentrated until the appropriate volume (typically from 10 to 20 μL). Peptides are purified by the ZipTipC18 columns (Millipore) according to the manufacturer's protocol and are mixed with matrix solution including αCHCA on the sample plate for MALDI-TOF/MS.
7. To identify proteins, we use 4,800 plus MALDI-TOF/TOF™ Analyzer (Applied Bioscience, USA), and results are analyzed by Protein Pilot™ software.

4. Notes

1. T-2 toxin should be stored at –20°C for preventing evaporation of ethanol, otherwise concentration of T-2 toxin will change.
2. If some detergents are used to dissolve protein samples, do not use ionized detergents such as SDS.
3. RC DC Protein Assay Kit is compatible with the following reagents and buffers: 2% CHAPS, 350 mM DTT, 0.1 M EDTA, 0.5 M imidazole, laemmli buffer with 5% β-mercaptoethanol, 2.5 M sodium hydrate, 2 mM TBP, 0.5 M Tris, 2% Triton, and 2% Tween-20.
4. Since primary amines, DTT, or ampholytes can react with NHS esters of the CyDyes, protein solutions do not have to include any of these substances.
5. DTT and iodoacetamide solutions should be made just before use.

References

1. de la Fuente van Bentem, S., Mentzen, W. I., de la Fuente, A., and Hirt, H. (2008) Towards functional phosphoproteomics by mapping differential phosphorylation events in signaling networks. *Proteomics* **8**, 4453–4465.
2. Lilley, K. S., and Dupree, P. (2006) Methods of quantitative proteomics and their application to plant organelle characterization. *J. Exp. Bot.* **57**, 1493–1499.
3. Unlü, M., Morgan, M. E., and Minden, J. S. (1997) Difference gel electrophoresis: A single gel method for detecting changes in protein extracts. *Electrophoresis* **18**, 2071–2077.

4. Timms, J. F., and Cramer, R. (2008) Difference gel electrophoresis. *Proteomics* **8**, 4886–4897.
5. Minden, J. S., Dowd, S. R., Meyer, H. E., and Stühler, K. (2009) Difference gel electrophoresis. *Electrophoresis* **30**, S156–S161.
6. Hrebicek, T., Dürrschmid, K., Auer, N., Bayer, K., and Rizzi, A. (2007) Effect of CyDye minimum labeling in differential gel electrophoresis on the reliability of protein identification. *Electrophoresis* **28**, 1161–1169.
7. Sudakin, D. L. (2003) Trichothecenes in the environment: Relevance to human health. *Toxicol. Lett.* **143**, 97–107.
8. Nishiuchi, T., Masuda, D., Nakashita, H., Ichimura, K., Shinozaki, K., Yoshida, S., Kimura, M., Yamaguchi, I., and Yamaguchi, K. (2006) Fusarium phytotoxin trichothecenes have an elicitor-like activity in *Arabidopsis thaliana*, but the activity differed significantly among their molecular species. *Mol. Plant Microbe Interact.* **19**, 512–520.

Index

Note: The letters 'f', 't' and 'n' following locators refer to figures, tables and notes respectively.

H. Kodama, A. Komamine (eds.), *RNAi and Plant Gene Function Analysis*, Methods in Molecular Biology 744, DOI 10.1007/978-1-61779-123-9, © Springer Science+Business Media, LLC 2011

MIX
Papier aus verantwortungsvollen Quellen
Paper from responsible sources
FSC® C105338

If you have any concerns about our products,
you can contact us on
ProductSafety@springernature.com

In case Publisher is established outside the EU,
the EU authorized representative is:
Springer Nature Customer Service Center GmbH
Europaplatz 3, 69115 Heidelberg, Germany

Printed by Libri Plureos GmbH
in Hamburg, Germany